Lecture Notes in Computer Science 15531

The series Lecture Notes in Computer Science (LNCS), including its subseries Lecture Notes in Artificial Intelligence (LNAI) and Lecture Notes in Bioinformatics (LNBI), has established itself as a medium for the publication of new developments in computer science and information technology research, teaching, and education.

LNCS enjoys close cooperation with the computer science R & D community, the series counts many renowned academics among its volume editors and paper authors, and collaborates with prestigious societies. Its mission is to serve this international community by providing an invaluable service, mainly focused on the publication of conference and workshop proceedings and postproceedings. LNCS commenced publication in 1973.

Violetta Weger · Jean-Christophe Deneuville ·
Anna-Lena Horlemann
Editors

Code-Based Cryptography

12th International Workshop, CBCrypto 2024
Zurich, Switzerland, May 25–26, 2024
Revised Selected Papers

Springer

Editors
Violetta Weger
Technical University of Munich
Munich, Germany

Jean-Christophe Deneuville
École Nationale de l'Aviation Civile
Toulouse, France

Anna-Lena Horlemann
University of St. Gallen
St. Gallen, Switzerland

ISSN 0302-9743 ISSN 1611-3349 (electronic)
Lecture Notes in Computer Science
ISBN 978-3-031-90228-4 ISBN 978-3-031-90229-1 (eBook)
https://doi.org/10.1007/978-3-031-90229-1

This Springer imprint is published by the registered company Springer Nature Switzerland AG
The registered company address is: Gewerbestrasse 11, 6330 Cham, Switzerland

Preface

Post-quantum cryptography has seen tremendous interest in recent years, especially after the first call for standardization of quantum-safe primitives by the US National Institute of Standards and Technology (NIST) back in 2017. Code-based cryptography, i.e., cryptographic systems based on algebraic codes, is arguably among the most relevant and active areas in post-quantum cryptography. This is clearly showcased in the additional call for post-quantum digital signatures from 2023, where 6 out of the 14 submissions are code-based.

Originally named the "Code-Based Cryptography (CBC) Workshop", the series was initiated in 2009 as an informal forum with the goal of bringing together researchers active in the analysis and development of code-based encryption and signature schemes. Over the years, the workshop has grown from a Europe-based, regional event to become the largest worldwide venue for the code-based cryptography community. The workshop was renamed "CBCrypto" in 2020, and its organization was co-located with the flagship conference Eurocrypt, and extended to include the publication of revised selected manuscripts in the form of a post-conference proceedings volume. Quickly, CBCrypto has become a popular event with high participation. The 2024 edition, with more than 100 participants, confirmed this trend.

This volume contains selected papers presented at CBCrypto 2024: The 12th International Workshop on Code-Based Cryptography, held in Zurich, Switzerland, during May 25–26, 2024. There were initially 41 submissions. Each submission was carefully reviewed by 2.3 people on average. Among these submissions, 15 were selected for a talk at the conference, while 4 of them were dedicated a slot in the poster session. Eventually, 9 papers were invited to submit a full version of their work, among which the committee decided to accept 5 papers.

The conference covered a wide range of important aspects of code-based cryptography such as cryptanalysis of existing schemes, the proposal of new cryptographic systems and protocols as well as improved decoding algorithms. As such, the works presented in this book provide a synthesized yet significant overview of the state of the art of code-based cryptography, laying out the groundwork for future developments.

We would like to thank the Program Committee members and the external reviewers for their hard and timely work. Additionally, we would like to thank the sponsors, TII, SCNAT, SMS and Infosec for their support.

December 2024

Violetta Weger
Jean-Christophe Deneuville
Anna-Lena Horlemann

Organization

Program Committee Chairs

Violetta Weger	Technical University of Munich, Germany
Jean-Christophe Deneuville	École Nationale de l'Aviation Civile, France
Anna-Lena Horlemann	University of St. Gallen, Switzerland

Program Committee

Marco Baldi	Università Politecnica delle Marche, Italy
Hannes Bartz	DLR, Germany
Emanuele Bellini	Technology Innovation Institute, UAE
Sergey Bezzateev	Saint Petersburg University of Airspace Instrumentation, Russia
Loïc Bidoux	Technology Innovation Institute, UAE
Maxime Bros	NIST, USA
Pierre-Louis Cayrel	Laboratoire Hubert Curien, France
Franco Chiaraluce	Università Politecnica delle Marche, Italy
Jean-Christophe Deneuville	École Nationale de l'Aviation Civile, France
Taraneh Eghlidos	Sharif University of Technology, Iran
Andre Esser	Technology Innovation Institute, UAE
Shay Gueron	University of Haifa, Israel and Intel Corporation, USA
Anna-Lena Horlemann	University of St. Gallen, Switzerland
Felice Manganiello	Clemson University, USA
Giacomo Micheli	University of South Florida, USA
Kirill Morozov	University of North Texas, USA
Gerardo Pelosi	Politecnico di Milano, Italy
Edoardo Persichetti	FAU, USA and Sapienza University of Rome, Italy
Joachim Rosenthal	University of Zurich, Switzerland
Simona Samardjiska	Radboud University, The Netherlands
Paolo Santini	Polytechnic University of Marche, Italy
Antonia Wachter-Zeh	Technical University of Munich, Germany
Violetta Weger	Technical University of Munich, Germany
Øyvind Ytrehus	University of Bergen, Norway

Additional Reviewers

Antognazza, Francesco
Aragon, Nicolas
Bitzer, Sebastian
Colombier, Brice
Karl, Patrick
Lichtinger, Jacob
Meneghetti, Alessio
Nugier, Cyrius
Pallozzi Lavorante, Vincenzo
Perlner, Ray
Ritterhoff, Stefan
Robert, Jean-Marc
Robinson, Angela
Slaughter, Freeman
Tognolini, Giovanni

Contents

HQC Beyond the BSC: Towards Error Structure-Aware Decoding

Marco Baldi[1], Sebastian Bitzer[2(✉)], Nicholas Lilla[1], and Paolo Santini[1]

[1] Università Politecnica delle Marche, Ancona, Italy
{m.baldi,p.santini}@univpm.it
[2] Technical University of Munich, Munich, Germany
sebastian.bitzer@tum.de

Abstract. In Hamming Quasi-Cyclic (HQC), one of the finalists in the NIST competition for the standardization of post-quantum cryptography, decryption relies on decoding a noisy codeword through a public error-correcting code. The noise vector has a special form that depends on the secret key (a pair of sparse polynomials). However, the decoder, which is currently employed in HQC, is agnostic to the secret key, operating under the assumption that the error arises from a Binary Symmetric Channel (BSC). In this paper, we demonstrate that this special noise structure can instead be leveraged to develop more powerful decoding strategies.

We first study the problem from a coding-theoretic perspective. The current code design, which admits a non-zero decryption failure rate, is close to optimal in the setting of a decoder that is agnostic to the error structure. We show that there are code-decoder pairs with a considerably shorter code length that can guarantee unique decoding by taking the error structure into account. This result is non-constructive, i.e., we do not provide an explicit code construction and it remains open whether efficient decoding is possible. Nevertheless, it highlights the potential that can be tapped by taking the error structure into account. We then argue that, in practice, the matter of decoding in HQC can be related to solving an instance of the noisy syndrome decoding problem, in which the parity-check matrix is constituted by the polynomials in the secret key. We show that, using decoders for Low-Density Parity-Check (LDPC) and Moderate-Density Parity-Check (MDPC) codes, one can significantly reduce the entity of the noise and, de facto, also the Decoding Failure Rate (DFR) of the HQC decoder.

This preliminary study leaves some open questions and problems. While it shows that decoding in HQC can be improved, the modeling of the DFR gets more complicated: even for the basic decoder we propose in this paper, we have not been able to devise a reliable DFR model. This is likely due to the fact that the decoder structure resembles the iterative nature of LDPC/MDPC decoders, for which devising a reliable DFR estimation is a well-known difficult problem.

Keywords: HQC · Decryption Failure Rate · Noisy Syndrome Decoding

V. Weger et al. (Eds.): CBCrypto 2024, LNCS 15531, pp. 1–24, 2025.
https://doi.org/10.1007/978-3-031-90229-1_1

1 Introduction

In code-based encryption schemes, the public key is usually a disguised representation of a secret error-correcting code $\mathscr{C}$ equipped with an efficient decoding algorithm. Recovering the structure of $\mathscr{C}$ shall be computationally infeasible: this is one of the security assumptions (the other being the hardness of decoding arbitrary linear codes). For instance, LEDAcrypt, BIKE, and Classic McEliece, some code-based cryptosystems participating in the NIST competition for the standardization of post-quantum cryptography [10], follow this paradigm: LEDAcrypt [5] relies on Low-Density Parity-Check (LDPC) codes, BIKE [4] exploits a special family of LDPC codes known as Moderate-Density Parity-Check (MDPC) codes, while Classic McEliece [2] employs binary Goppa codes.

HQC [9], another alternate in the NIST competition, departs from such a paradigm: no secret error-correcting code is required. Instead, a publicly known code $\mathscr{C}$, equipped with an efficient decoder Dec, is employed. The possibility to perform efficient decoding is based on the fact that the legitimate receiver, using the secret key, can perform some efficient manipulations of the ciphertext (essentially, a couple of polynomial multiplications and sums) resulting in $\mathbf{c} = \mathbf{t} + \mathbf{z}$, with $\mathbf{t} \in \mathscr{C}$ and

$$\mathbf{z} = \mathbf{x} \cdot \mathbf{r}^{(2)} + \mathbf{y} \cdot \mathbf{r}^{(1)} + \mathbf{e}. \tag{1}$$

All terms in the previous formula are sparse polynomials in the ring $\mathbb{F}_2[X]/(X^n - 1)$; hence $\mathbf{z}$ is sparse as well. In particular, $\mathbf{x}$ and $\mathbf{y}$ constitute the secret key while $\mathbf{r}^{(1)}$, $\mathbf{r}^{(2)}$ and $\mathbf{e}$ are ephemeral and are chosen uniformly at random by the sender. Decoding $\mathbf{c}$ through the public code $\mathscr{C}$ allows to correct $\mathbf{z}$.

The error $\mathbf{z}$ is modeled through a Binary Symmetric Channel (BSC); the code $\mathscr{C}$ is chosen accordingly, and, in principle, any error-correcting code $\mathscr{C}$ can be used. In all four rounds of the NIST competition, the authors of HQC have selected $\mathscr{C}$ as the tensor product between two error-correcting codes. This results in a good trade-off between computational efficiency, error-correction capability and the possibility to derive a solid and reliable Decoding Failure Rate (DFR) modeling. We remark that the availability of such a model is crucial, as IND-CCA2 security can only be achieved if the DFR is negligible in the security parameter.

1.1 Our Contribution

The decoder employed in HQC is agnostic to the secret key, i.e., it never explicitly takes advantage of the particular structure (1) of the error vector $\mathbf{z}$. However, the secret-key polynomials $\mathbf{x}$ and $\mathbf{y}$ are part of how $\mathbf{z}$ is computed: as we show in this paper, this information can be employed to devise more powerful decoding strategies.

Coding-Theory Perspective. First, with coding theory arguments, we show that taking the structure of $\mathbf{z}$ and the knowledge of $\mathbf{x}$ and $\mathbf{y}$ into account enables shorter code lengths. To this end, we first revisit fundamental bounds on the required code length that are implied by modeling errors as caused by a BSC. These bounds show that the code construction used in the round-4 version of HQC achieves a shorter length than would be possible if correct decryption has to be guaranteed. On the other hand, an arbitrary code-decoder pair cannot do considerably better than the current choice, that is, as long as the code design and the decoding algorithm follow the assumption that errors are caused by a BSC.

The situation changes when we consider the exact structure of $\mathbf{z}$ and make the decoder aware of $\mathbf{x}$ and $\mathbf{y}$. Using an argument similar to the classical Gilbert-Varshamov (GV) bound, we show that codes able to guarantee correct decryption exist with significantly shorter lengths than under the BSC model. As is typical for GV-like arguments, this result is not constructive, i.e., we do not provide an explicit code construction or an efficient decoding algorithm.

The Relation with Noisy-Syndrome Decoding. When $\mathscr{C}$ is a tensor product code, the problem of recovering $\mathbf{r}^{(1)}$ and $\mathbf{r}^{(2)}$ can be formulated as a noisy syndrome decoding problem [6]. Indeed, decoding the inner code yields an estimate $\widehat{\mathbf{z}}$ for $\mathbf{z}$. We view $\widehat{\mathbf{z}}$ as a noisy syndrome, as it can be expressed as

$$\widehat{\mathbf{z}} = \underbrace{\left(\mathrm{Rot}(\mathbf{x})||\mathrm{Rot}(\mathbf{y})\right)}_{\mathbf{H}}\mathbf{r}^\top + \varDelta\mathbf{z} = \mathbf{H}\mathbf{r}^\top + \varDelta\mathbf{z},$$

where $\mathrm{Rot}(\mathbf{x})$ and $\mathrm{Rot}(\mathbf{y})$ are the circulant matrices whose first rows are $\mathbf{x}$ and $\mathbf{y}$, respectively, while $\mathbf{r}$ is the length-$2n$ vector formed by the coefficients of the unknown polynomials $\mathbf{r}^{(1)}$ and $\mathbf{r}^{(2)}$. The product $\mathbf{H}\mathbf{r}^\top$ is a syndrome affected by the noise contained in $\varDelta\mathbf{z}$. The amount of noise depends on the quality of decoding: the lower the number of decoding errors, the smaller the weight of $\varDelta\mathbf{z}$.

Since $\mathbf{x}$ and $\mathbf{y}$ are sparse polynomials, $\mathbf{H}$ is sparse as well. Then, $\mathbf{H}$ can be interpreted as the parity-check matrix of a Low-Density Parity-Check (LDPC) or Moderate-Density Parity-Check (MDPC) code, and $\mathbf{r}$ can be estimated using decoders for such codes. Note that these algorithms, such as the Bit Flipping decoder, naturally tolerate a syndrome affected by a moderate amount of noise.

Let $\widehat{\mathbf{r}}^{(1)}$ and $\widehat{\mathbf{r}}^{(2)}$ be the obtained estimates: they can be used to reduce the noise affecting $\mathbf{c}$, as one can update $\mathbf{c}$ as

$$\begin{aligned}\widehat{\mathbf{c}} &= \mathbf{c} - \mathbf{x}\cdot\widehat{\mathbf{r}}^{(2)} - \mathbf{y}\cdot\widehat{\mathbf{r}}^{(1)}\\ &= \mathbf{t} + \mathbf{z} - \mathbf{x}\cdot\widehat{\mathbf{r}}^{(2)} - \mathbf{y}\cdot\widehat{\mathbf{r}}^{(1)}\\ &= \mathbf{t} + \mathbf{x}\cdot\underbrace{\left(\mathbf{r}^{(2)} - \widehat{\mathbf{r}}^{(2)}\right)}_{\varDelta\mathbf{r}^{(2)}} + \mathbf{y}\cdot\underbrace{\left(\mathbf{r}^{(1)} - \widehat{\mathbf{r}}^{(1)}\right)}_{\varDelta\mathbf{r}^{(1)}} + \mathbf{e} = \mathbf{t} + \underbrace{\mathbf{x}\cdot\varDelta\mathbf{r}^{(2)} + \mathbf{y}\cdot\varDelta\mathbf{r}^{(1)} + \mathbf{e}}_{\mathbf{z}^*}.\end{aligned}$$

The better the estimates $\widehat{\mathbf{r}}^{(1)}$ and $\widehat{\mathbf{r}}^{(2)}$, the lower the weights of $\varDelta\mathbf{r}^{(1)}$ and $\varDelta\mathbf{r}^{(2)}$ and, consequently, the lower the weight of $\mathbf{z}^*$. In particular, if the weights of $\varDelta\mathbf{r}^{(1)}$ and $\varDelta\mathbf{r}^{(2)}$ are lower than the weights of $\mathbf{r}^{(1)}$ and $\mathbf{r}^{(2)}$, then with high probability the noise affecting $\mathbf{t}$ is reduced.

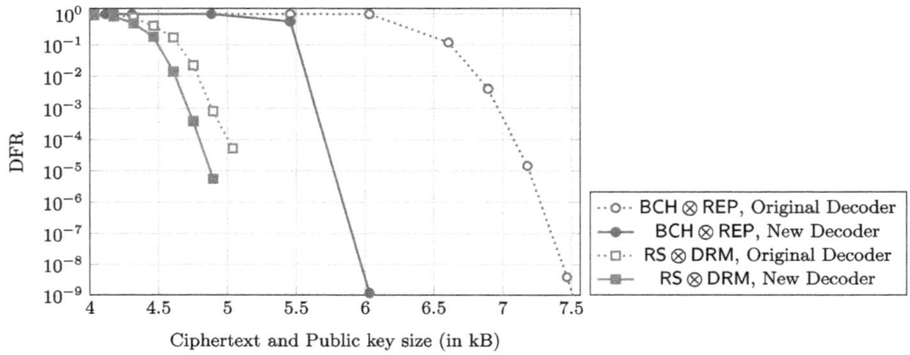

Fig. 1. DFR vs ciphertext plus public key sizes. The DFR has been estimated using numerical simulations by varying the lengths of the component codes. The ciphertext has length $\approx 2n$, while the public key has length $\approx n$.

In this paper, we consider a straightforward decoding strategy, which can be seen as a single iteration of the Bit Flipping (BF) decoder. We show (both theoretically as well as through numerical simulations) that, even with this simple algorithm, one can obtain pretty good estimates $\widehat{\mathbf{r}}^{(1)}$ and $\widehat{\mathbf{r}}^{(2)}$.

Reducing the DFR. After the initial filtering step, one can proceed by applying the standard decoder of $\mathscr{C}$ on $\widehat{\mathbf{c}}$. This is already sufficient to improve decoding significantly so that shorter codes can be used to achieve the desired DFR: this reduces both the ciphertext and public key size. In Fig. 1 we show an example of the resulting gains, considering the choices for $\mathscr{C}$ which have been used by HQC in the NIST competition: the product of a BCH and a Repetition code (which we indicate as $\mathsf{BCH} \otimes \mathsf{REP}$), and the product of a Reed-Solomon and a Duplicated Reed-Muller code (which we indicate as $\mathsf{RS} \otimes \mathsf{DRM}$).

Limitations and Open Questions. This work leaves some open questions. First, many different techniques exist to estimate $\mathbf{r}^{(1)}$ and $\mathbf{r}^{(2)}$ as, in principle, any LDPC/MDPC decoder may be used to decode the noisy syndrome. Each of these techniques comes with different trade-offs: defining the best strategy to exploit knowledge of $\mathbf{x}$ and $\mathbf{y}$ is beyond the scope of this paper.

Secondly, and more importantly, one of the key advantages of HQC is the possibility to derive a reliable and closed-form formula for the DFR. While more involved decoders may improve error-correction capability and overall performance, they also come with a more complex DFR analysis. For instance, even for the simple decoder we present in this paper, we have not been able to derive a reliable theoretical model for the DFR. This is due to some technical caveats that arise when iterative decoders (such as the one we propose in this paper) are considered. In our view, this is the most important open question: can we design a decoder with a lower DFR while also allowing for a reliable DFR prediction?

1.2 Paper Organization

The paper is organized as follows. In Sect. 2, we establish the notation used throughout the paper and recall the main properties of HQC. In Sect. 3, we provide coding-theoretic arguments that support the possibility of improving performance. Section 4 describes how HQC decoding relates to a noisy syndrome decoding problem and analyzes the use of the BF decoder in such a context. The impact on the DFR of HQC is discussed in Sect. 5, while Sect. 6 ends the paper with some concluding remarks.

2 Notation and Background

We use standard notations for finite fields: for q a prime power, $\mathbb{F}_q$ denotes a finite field with q elements while $\mathbb{F}_q[X]$ is the ring of polynomials with coefficients in $\mathbb{F}_q$. For the majority of the paper, we work with the binary finite field $\mathbb{F}_2$. For an integer n, we denote $\mathcal{R} := \mathbb{F}_2[X]/(X^n-1)$. We frequently view the elements of $\mathcal{R}$ as length-n vectors over $\mathbb{F}_2$, relying on the following canonical representation:

$$\sum_{i=1}^{n} a_i X^{i-1} = \mathbf{a} \in \mathcal{R} \iff (a_1, a_2, \cdots, a_n) = \mathbf{a} \in \mathbb{F}_2^n.$$

For a polynomial $\mathbf{a} \in \mathcal{R}$, we indicate by $\mathrm{wt}(\mathbf{a})$ its Hamming weight, i.e., the number of non-zero coefficients. By support of a polynomial, which we indicate as $\mathrm{supp}(\mathbf{a})$, we refer to the set of indices i for which $a_i \neq 0$; notice that $\mathrm{wt}(\mathbf{a}) = |\mathrm{supp}(\mathbf{a})|$. Given two polynomials $\mathbf{a}$ and $\mathbf{b}$, we indicate the j-th coefficient of their product as $(\mathbf{a} \cdot \mathbf{b})_j$. For $w \in \mathbb{N}$, $w \leqslant n$, we define $\mathcal{R}_w = \{\mathbf{a} \in \mathcal{R} \mid \mathrm{wt}(\mathbf{a}) = w\}$.

Probability Distributions. Given a set A, we write $a \xleftarrow{\$} A$ when a is drawn uniformly at random from A. We use $\mathcal{B}_{n,\rho}$ to indicate the Bernoulli distribution over $\mathbb{F}_2^n$ with parameter ρ, i.e., the distribution that returns vectors of length n and such that any entry is 1 with probability ρ and 0 with probability $1-\rho$. If $\mathbf{a} \in \mathbb{F}_2^n$ is distributed according to $\mathcal{B}_{n,\rho}$, we write $\mathbf{a} \sim \mathcal{B}_{n,\rho}$.

2.1 Background on Coding Theory

A linear code $\mathscr{C} \subseteq \mathbb{F}_q^n$ with length n and dimension k is a linear subspace of $\mathbb{F}_q^n$ containing q^k vectors, called *codewords*. A compact representation for a code is a *generator matrix*, that is, a matrix $\mathbf{G} \in \mathbb{F}_q^{k\times n}$ with rank k, whose row span yields $\mathscr{C}$, that is, $\mathscr{C} = \{\mathbf{mG} \mid \mathbf{m} \in \mathbb{F}_q^k\}$. The minimum distance of a code is the minimum Hamming distance between two distinct codewords and, for linear codes, corresponds to the minimum weight of a non-zero codeword.

We say $\mathscr{C}$ is an *error-correcting code* whenever it can be equipped with an efficient algorithm Dec that, on input some $\mathbf{c} = \mathbf{mG} + \mathbf{e}$, returns $\mathbf{mG}$ with high probability when $\mathbf{e}$ has sufficiently low weight. When the decoder does not return $\mathbf{mG}$, we encounter a *decoding failure*.[1] The probability (over the channel's and

[1] Generically, decoding failures happen when the decoder either halts on a vector that is not a codeword or on a codeword different from the transmitted one.

decoder's randomness) that a decoding failure event occurs is called the Decoding Failure Rate (DFR).

Tensor Product Codes. Let q_1 and q_2 be two (possibly equal) prime powers. Given two codes $\mathscr{C}_1 \subseteq \mathbb{F}_{q_1}^{n_1}$ and $\mathscr{C}_2 \subseteq \mathbb{F}_{q_2}^{n_2}$ with dimensions k_1 and k_2, we denote by $\mathscr{C} = \mathscr{C}_1 \otimes \mathscr{C}_2$ their *tensor product*. Sometimes, $\mathscr{C}_1$ is called *outer code* and $\mathscr{C}_2$ *inner code*. Let $\mathsf{Enc}_1 : \mathbb{F}_{q_1}^{k_1} \mapsto \mathscr{C}_1$ and $\mathsf{Enc}_2 : \mathbb{F}_{q_2}^{k_2} \mapsto \mathscr{C}_2$ be the encoding functions for $\mathscr{C}_1$ and $\mathscr{C}_2$. Analogously, we denote by Dec_1 and Dec_2 their decoding algorithms. Finally, we denote by $\mathsf{Enc}_{\mathbb{F}_{q_1},\mathbb{F}_{q_2}}$ a function that, on input a vector in $\mathbb{F}_{q_1}^{n_1}$, returns a sequence with values over $\mathbb{F}_{q_2}$ and length n_1'; we require that n_1' is a multiple of k_2.[2] The inverse function is indicated as $\mathsf{Dec}_{\mathbb{F}_{q_1},\mathbb{F}_{q_2}}$. If the two finite fields are the same, the function $\mathsf{Enc}_{\mathbb{F}_{q_1},\mathbb{F}_{q_2}}$ is simply the identity (and so is its inverse). Thus, we also have $n_1' = n_1$ hence k_2 must divide n_1.

We now show how encoding of a message $\mathbf{m} \in \mathbb{F}_q^{k_1}$ works. First, $\mathbf{m}$ is encoded into a codeword of $\mathscr{C}_1$. Then, we apply $\mathsf{Enc}_{\mathbb{F}_{q_1},\mathbb{F}_{q_2}}$ to obtain a string of length n_1' and values over $\mathbb{F}_{q_2}$. This is divided into n_1'/k_2 chunks: each has length k_2 and is encoded into a codeword of $\mathscr{C}_2$. The resulting codeword has length $n_1'n_2/k_2$. See Fig. 2 for a graphical representation of the encoding procedure.

Decoding a tensor product code is done by first decoding the $\frac{n_1'}{k_2}$ codewords of $\mathscr{C}_2$ and then decoding their concatenation through $\mathscr{C}_1$. Namely, let $\mathbf{c} = \left(\mathbf{c}^{(1)}||\cdots||\mathbf{c}^{(n_1'/k_2)}\right)$ be a received word, with each $\mathbf{c}^{(i)}$ having length n_2 (here, we use $||$ to indicate the concatenation of two row vectors). Then, decoding of $\mathbf{c}$ is done by first decoding each $\mathbf{c}^{(i)}$ and mapping the result back to $\mathbb{F}_{q_1}$ through $\mathsf{Dec}_{\mathbb{F}_{q_1},\mathbb{F}_{q_2}}$; all these words are concatenated and the obtained word is decoded through $\mathscr{C}_1$. Compactly, this process is described as

$$\mathsf{Dec}_1\left(\mathsf{Dec}_{\mathbb{F}_{q_1},\mathbb{F}_{q_2}}\left(\mathsf{Dec}_2(\mathbf{c}^{(1)})\right)\Big|\Big|\cdots\Big|\Big|\mathsf{Dec}_{\mathbb{F}_{q_1},\mathbb{F}_{q_2}}\left(\mathsf{Dec}_2(\mathbf{c}^{(n_1'/k_2)})\right)\right).$$

2.2 HQC in a Nutshell

Figure 3 summarizes the main operations in HQC. Given the scope of this paper, we omit all the operations that are not relevant from a coding-theory perspective (but, obviously, are relevant from a cryptographic perspective!), such as the use of seeds and the IND-CCA2 conversion; for full details, we refer the interested reader to [9].

In our description, the encoding and decoding processes of the public code $\mathscr{C}$ are indicated as Enc and Dec, respectively. The outer code is generically defined over a finite field of size q, while the inner code is always binary.

Due to the special structure of the ciphertext and public key, one has $\mathbf{c} = \mathbf{t}+\mathbf{z}$, where $\mathbf{t} \in \mathscr{C}$ and

$$\mathbf{z} = \mathbf{x} \cdot \mathbf{r}^{(2)} + \mathbf{y} \cdot \mathbf{r}^{(1)} + \mathbf{e}. \tag{2}$$

Since all the polynomials in the above equation are sparse, $\mathbf{z}$ has low weight. This allows for efficiently correcting errors through the decoding algorithm Dec.

[2] For instance, if $q_1 = 2^\ell$ and $q_2 = 2$, a codeword of $\mathscr{C}_1$ can be represented as a binary string of length $\ell \cdot n_1$.

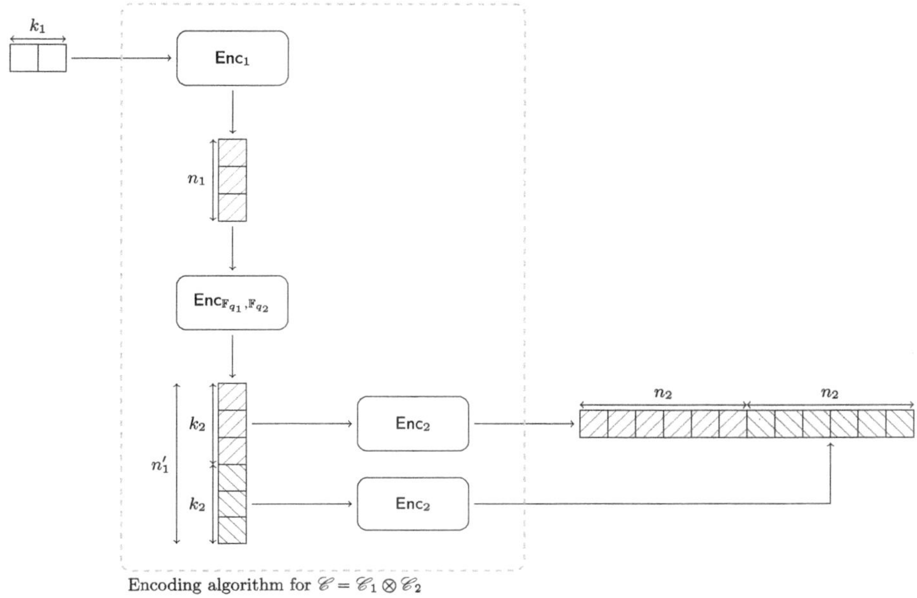

Fig. 2. Example of the encoding procedure for a tensor code obtained from two component codes defined over two fields $\mathbb{F}_{q_1}$ and $\mathbb{F}_{q_2}$ such that $q_1 = q_2^2$. The function $\mathsf{Enc}_{\mathbb{F}_{q_1},\mathbb{F}_{q_2}}$ maps a symbol from $\mathbb{F}_{q_1}$ into two symbols of $\mathbb{F}_{q_2}$. Code parameters are $k_1 = 2$, $n_1 = 3$, $n'_1 = 2 \cdot n_1 = 6$, $k_2 = 3$ and $n_2 = 6$. The resulting code has dimension $k = k_1 = 2$ and $n = \frac{n'_1}{k_2} \cdot n_2 = 12$.

Choice of $\mathscr{C}$. The practical performance of the scheme heavily depends on the choice of $\mathscr{C}$. Indeed, one should choose a family of error-correcting codes that yields a good trade-off between error-correction capability, compactness, and computational efficiency. First, notice that both the public key size and the ciphertext size are linear in n.[3] Hence, one would desire n to be as small as possible; however, this value cannot be too small as we must guarantee that, with sufficiently large probability, Dec corrects (efficiently) the errors in $\mathbf{z}$.

To guarantee IND-CCA2 security, the DFR must be negligible in the security parameter, i.e., less than $2^{-\lambda}$. This leads to another requirement: devising a theoretical, closed-form formula for the DFR must be feasible. For this reason, the decoding process cannot be too involved: as it is well known, devising reliable models for the DFR is extremely complicated, and only a few decoding algorithms (namely, the simplest ones) allow for it.

[3] The public key size is $n + \lambda$ bits, with λ denoting the security level, because $\mathbf{h}$ can be compressed by the seed with which it has been generated. The ciphertext size, instead, corresponds to $2n + 2\lambda$. The term 2λ is the overhead due to the IND-CCA2 conversion.

Key generation:
1. sample $\mathbf{h} \xleftarrow{\$} \mathcal{R}$;
2. sample $\mathbf{x}, \mathbf{y} \xleftarrow{\$} \mathcal{R}_w \times \mathcal{R}_w$;
3. set $\mathsf{sk} := (\mathbf{x}, \mathbf{y})$ and $\mathsf{pk} := \{\mathbf{h}\ ,\ \mathbf{s} = \mathbf{x} + \mathbf{h} \cdot \mathbf{y}\}$.

Encryption: on input $\mathbf{m} \in \mathbb{F}_q^k$ and $\mathsf{pk} := \{\mathbf{h}, \mathbf{s}\}$, do:
1. sample $\mathbf{r}^{(1)}, \mathbf{r}^{(2)} \xleftarrow{\$} \mathcal{R}_{w_r} \times \mathcal{R}_{w_r}$, $\mathbf{e} \xleftarrow{\$} \mathcal{R}_{w_e}$;
2. set $\mathbf{u} = \mathbf{r}^{(1)} + \mathbf{h} \cdot \mathbf{r}^{(2)}$;
3. compute $\mathbf{t} = \mathsf{Enc}(\mathbf{m})$;
4. set $\mathbf{v} = \mathbf{t} + \mathbf{e} + \mathbf{s} \cdot \mathbf{r}^{(2)}$;
5. the ciphertext is $\{\mathbf{u}, \mathbf{v}\}$.

Decryption: on input $(\mathbf{u}, \mathbf{v}) \in \mathcal{R}^2$ and $\mathsf{sk} := (\mathbf{x}, \mathbf{y})$, do:
1. compute $\mathbf{c} = \mathbf{v} + \mathbf{y} \cdot \mathbf{u}$;
2. run $\mathsf{Dec}(\mathbf{c})$.

Fig. 3. Description of HQC.

Taking this into account, the authors of HQC chose to use tensor product codes, as they allow for a simple DFR modeling and, at the same time, offer a good error-correction capability. While the design of HQC has essentially remained the same throughout the years, its parameters and the choice of $\mathscr{C}$ have changed. Two choices for $\mathscr{C}$ have been considered so far:

- Version $\mathsf{BCH} \otimes \mathsf{Rep}$ from first round:
  - Outer code: BCH code defined over $\mathbb{F}_2$
  - Inner code: Repetition code defined over $\mathbb{F}_2$
- Version $\mathsf{RS} \otimes \mathsf{DRM}$ from second round:
  - Outer code: Reed-Solomon code defined over $\mathbb{F}_{2^8}$
  - Inner code: Duplicated Reed-Muller code defined over $\mathbb{F}_2$

3 Coding-Theoretic Analysis

This section highlights from a coding-theoretic perspective the improvements that are possible due to the knowledge of the error structure at the decoder (in particular due to knowing $\mathbf{x}, \mathbf{y}$). To this end, we first revisit decoding agnostic to the secret key and its limitations. Then, we derive a suitable generalization of the Gilbert-Varshamov (GV) bound tailored to the particular setting we encounter in HQC. For simplicity, throughout this section, we assume that the used code is of the same length as the vectors in $\mathcal{R}$.

3.1 Approximating HQC Errors via a Binary Symmetric Channel

The HQC error being of the form $\mathbf{z} = \mathbf{x} \cdot \mathbf{r}^{(2)} + \mathbf{y} \cdot \mathbf{r}^{(1)} + \mathbf{e}$ implies that

$$\mathrm{wt}(\mathbf{z}) \leqslant 2 \cdot w_r \cdot w + w_e := w_{\max}.$$

Besides the maximum weight of $\mathbf{z}$, the probability that a single coordinate of $\mathbf{z}$ is non-zero over the choice of $\mathbf{r}^{(1)}, \mathbf{r}^{(2)}, \mathbf{e}$ can be computed as

$$\Pr[z_i = 1] = 2\rho'_{w,w_r}(1-\rho_{w,w_r})\left(1-\frac{w_e}{n}\right) + \left((1-\rho_{w,w_r})^2 + \rho^2_{w,w_r}\right)\frac{w_e}{n} := \rho_z$$

where $\rho_{w,w_r} = \binom{n}{w}^{-1}\sum_{\ell \text{ odd}}\binom{w_r}{\ell}\binom{n-w_r}{w-\ell}$ [3]. For completeness, the proof is included in Appendix A. Deriving further statements beyond the maximum weight and the error probability of a single coordinate is involved. Therefore, the following heuristic is used to derive an analytical expression for the DFR.

Heuristic 1 (Binomial Approximation [3]). *Under the simplifying assumption that the coordinates of* $\mathbf{z}$ *are independent random variables, the probability of* $\mathbf{z}$ *is modeled as*

$$\Pr[\mathbf{z}] = \begin{cases} N \cdot \rho_z^{\mathrm{wt}(\mathbf{z})} \cdot (1-\rho_z)^{n-\mathrm{wt}(\mathbf{z})} & \textit{if } \mathrm{wt}(\mathbf{z}) \leqslant \min\{w_{\max}, n\}, \\ 0 & \textit{else,} \end{cases}$$

with normalization factor $N = \Pr\left[\mathrm{wt}(z) > w_{\max} \mid \mathcal{B}_{n,\rho_z}\right]^{-1}$*. In particular, the weight* $\mathrm{wt}(\mathbf{z})$ *follows a truncated binomial distribution.*

Heuristic 1 implies that HQC can operate with DFR $= 0$ only at the cost of a severe rate restriction. The following lemma emphasizes this.

Lemma 1 (DFR = 0 under Heuristic *1*).
Under Heuristic 1, any code-decoder pair that can guarantee correct decryption satisfies a code size of

$$|\mathscr{C}| \leqslant \frac{2^n}{\sum_{i=0}^{w_{\max}}\binom{n}{i}}.$$

Proof. To guarantee correct decoding under Heuristic 1, $\mathscr{C}$ needs to be able to correct all patterns of weight at most $w_{\max}$. The classical Hamming bound implies $|\mathscr{C}| \leqslant 2^n / \sum_{i=0}^{w_{\max}}\binom{n}{i}$. □

Note that the cryptographic setting determines the code dimension of the public code; for example, in the round-4 version of HQC, $k = \lambda$ is picked. In this setting, Lemma 1 can be understood as a lower bound on the required code length, which, in turn, determines the sizes of the public key and ciphertext.

In practice, reduced sizes can be achieved by tolerating a non-zero DFR. IND-CCA2 security can be guaranteed for a DFR of at most $2^{-\lambda}$. The following theorem, inspired by [7], gives a lower bound on the DFR for a given code size and dimension. Note that a related bound has been developed in [1].

Lemma 2 (Elias sphere-packing bound under Heuristic *1*).
Let $\mathscr{C}$ *be an arbitrary linear code of length* n *and dimension* k*. Denote as* ρ_z *the error probability of the BSC as in Heuristic 1. Let* $t \in [n]$ *be minimal such that* $\sum_{i=0}^{t}\binom{n}{i} > 2^{n-k}$*. Then, any decoder for* $\mathscr{C}$ *encounters at least a DFR of*

$$\sum_{i=t+1}^{w_{\max}}\binom{n}{i}\rho_z^i(1-\rho_z)^{n-i}.$$

Proof. Denote as $\mathcal{D}_{\mathscr{C}} \subset \mathbb{F}_2^n$ the decoding region of the linear code $\mathscr{C}$, i.e., $\mathbf{t} = \mathsf{Enc}(\mathbf{t} + \mathbf{z})$ for all $\mathbf{z} \in \mathcal{D}_{\mathscr{C}}$ and all $\mathbf{t} \in \mathscr{C}$. Note that $|\mathcal{D}_{\mathscr{C}}| = 2^{n-k}$. Then, since $\Pr[\mathbf{z}_1] > \Pr[\mathbf{z}_2]$ for $\mathrm{wt}(\mathbf{z}_1) < \mathrm{wt}(\mathbf{z}_2)$,

$$\epsilon = \Pr[\mathbf{z} \notin \mathcal{D}_{\mathscr{C}}] \geq \Pr[\mathbf{z} \notin \mathcal{B}_{\mathscr{C}}],$$

where $\mathcal{B}_{\mathscr{C}}$ is a Hamming ball with $|\mathcal{B}_{\mathscr{C}}| \geq |\mathcal{D}_{\mathscr{C}}|$. □

Again, by fixing $k = \lambda$ and a sufficiently low DFR, Lemma 2 can be understood as a lower bound on the required code length. The following example illustrates how the bounds compare with the code construction used in HQC.

Example 1. The concatenation of the Reed-Muller code and the Reed-Solomon code employed in round 4 by HQC uses $n = 17996$ for a security level of $\lambda = 128$ bit. The code dimension is $k = \lambda$, and the achieved DFR is at most $2^{-\lambda}$. Lemma 2 implies that the minimum code length for achieving the required DFR is at least $n = 13438$. Achieving guaranteed correctness of the decoding would require at least $n = 21822$ according to Lemma 1.

Example 1 shows that the current HQC parameters are already close to optimal, assuming that HQC errors follow Heuristic 1. This does not hold once we replace the binomial approximation with the actual structure of the HQC errors, as we will see in the following.

3.2 Beyond the BSC: Leveraging the Error Structure, x, and y

Let us denote the set of all possible error patterns given $\mathbf{x}, \mathbf{y}$ as

$$\mathcal{E}_{\mathbf{x},\mathbf{y}} = \left\{ \mathbf{z} = \mathbf{x} \cdot \mathbf{r}^{(2)} + \mathbf{y} \cdot \mathbf{r}^{(1)} + \mathbf{e} \mid \mathrm{wt}(\mathbf{r}^{(1)}) = w_r, \mathrm{wt}(\mathbf{r}^{(2)}) = w_r, \mathrm{wt}(\mathbf{e}) = w_r \right\}.$$

Then, the set of all possible error patterns for arbitrary $\mathbf{x}, \mathbf{y}$ is given by

$$\mathcal{E} = \bigcup_{\mathbf{x},\mathbf{y}:\ \mathrm{wt}(\mathbf{x}) = \mathrm{wt}(\mathbf{y}) = w} \mathcal{E}_{\mathbf{x},\mathbf{y}}.$$

There is a well-known condition for characterizing whether a code can uniquely correct all patterns in $\mathcal{E}$; see, e.g., [8].

Lemma 3. (Generalized GV bound).
A linear code $\mathscr{C}$ can correct all patterns in $\mathcal{E}$ uniquely if and only if $\mathscr{C} \cap \Delta\mathcal{E} = \{\mathbf{0}\}$, where $\Delta\mathcal{E}$ denotes the difference set of $\mathcal{E}$, i.e., $\Delta\mathcal{E} := \{\mathbf{z} - \mathbf{z}' \mid \mathbf{z}, \mathbf{z}' \in \mathcal{E}\}$. As a consequence, there exists a code of cardinality

$$|\mathscr{C}| \leqslant \frac{2^n}{|\Delta\mathcal{E}|}$$

that can correct all errors in $\mathcal{E}$.

Note that Lemma 3 generalizes the standard Hamming-metric Gilbert-Varshamov bound to arbitrary error patterns. We obtain the following lemma for our particular set consisting of HQC errors.

Lemma 4. (GV-like bound with error structure)
Let $\mathcal{E}$ be the set containing all possible HQC error patterns generated with parameters w and w_r. Then, $|\Delta\mathcal{E}| \leqslant \binom{n}{w}^4 \binom{n}{w_r}^6$ which implies that there exists a code $\mathscr{C}$ of dimension

$$k \geq n - 4w\log_2(\tfrac{n \cdot e}{w}) - 6w_r \log_2(\tfrac{n \cdot e}{w_r})$$

that can correct arbitrary HQC errors, i.e., guarantee correct decryption.

Proof. $\Delta\mathcal{E}$ contains vectors of the form

$$\mathbf{x} \cdot \mathbf{r}^{(2)} + \mathbf{y} \cdot \mathbf{r}^{(1)} + \mathbf{e} + \mathbf{x}' \cdot \mathbf{r}^{(2)'} + \mathbf{y}' \cdot \mathbf{r}^{(1)'} + \mathbf{e}'$$

with $\mathrm{wt}(\mathbf{x}) = \mathrm{wt}(\mathbf{x}') = \mathrm{wt}(\mathbf{y}) = \mathrm{wt}(\mathbf{y}') = w$, $\mathrm{wt}(\mathbf{r}^{(1)}) = \mathrm{wt}(\mathbf{r}^{(1)'}) = \mathrm{wt}(\mathbf{r}^{(2)}) = \mathrm{wt}(\mathbf{r}^{(2)'})$ $\mathrm{wt}(\mathbf{e}) = \mathrm{wt}(\mathbf{e}') = w_r$. The number of elements is upper-bounded by the number of choices for the components. This argument implies

$$|\mathcal{E}| \leqslant \binom{n}{w}^4 \binom{n}{w_r}^6.$$

Using the relation $\binom{a}{b} \leqslant \left(\frac{a \cdot e}{b}\right)^b$, the result follows via Lemma 3. □

Lemma 4 considers the particular structure of HQC errors but neglects that the decoder can use the knowledge of $\mathbf{x}, \mathbf{y}$. In the following, we analyze the impact of this additional information. To this end, let us first consider the case that the code $\mathscr{C}$ is only supposed to work for a particular choice of $\mathbf{x}, \mathbf{y}$. Then, $\mathscr{C} \cap \Delta\mathcal{E}_{\mathbf{x},\mathbf{y}} = \{\mathbf{0}\}$ is sufficient. For HQC, the public code is fixed and independent of the choice of the private key. Hence, the instantiation of $\mathscr{C}$ is supposed to work for all possible choices of $\mathbf{x}, \mathbf{y}$. Nevertheless, as the following lemma shows, knowing $\mathbf{x}, \mathbf{y}$ is valuable information for the decoder.

Lemma 5. *A code $\mathscr{C}$ with a decoder knowing $\mathbf{x}, \mathbf{y}$ can correct arbitrary HQC errors if and only if $\mathscr{C} \cap \Delta\mathcal{E}^* = \{\mathbf{0}\}$, where*

$$\Delta\mathcal{E}^* := \bigcup_{\mathbf{x},\mathbf{y}:\mathrm{wt}(\mathbf{x})=\mathrm{wt}(\mathbf{y})=w} \Delta\mathcal{E}_{\mathbf{x},\mathbf{y}}.$$

Proof. We require that $\mathscr{C} \cap \Delta\mathcal{E}_{\mathbf{x},\mathbf{y}} = \{\mathbf{0}\}$ for all $\mathbf{x}$ and $\mathbf{y}$. This is equivalent to

$$\bigcup_{\mathbf{x},\mathbf{y}:\mathrm{wt}(\mathbf{x})=\mathrm{wt}(\mathbf{y})=w} (\mathscr{C} \cap \Delta\mathcal{E}_{\mathbf{x},\mathbf{y}}) = \mathscr{C} \cap \bigcup_{\mathbf{x},\mathbf{y}:\mathrm{wt}(\mathbf{x})=\mathrm{wt}(\mathbf{y})=w} \Delta\mathcal{E}_{\mathbf{x},\mathbf{y}} = \{0\}.$$

□

The restriction on $\mathscr{C}$ due to Lemma 5 is less strict than the restriction implied by ignoring $\mathbf{x}, \mathbf{y}$, i.e., $\Delta\mathcal{E}^* \subseteq \Delta\mathcal{E}$. To see this, observe that $\Delta\mathcal{E}^*$ contains all vectors of the shape

$$\mathbf{r}^{(2)'} \cdot \mathbf{x} + \mathbf{r}^{(1)'} \cdot \mathbf{y} + \mathbf{e}$$

with $\mathrm{wt}(\mathbf{x}) = \mathrm{wt}(\mathbf{y}) = w$ and $\mathrm{wt}(\mathbf{r}^{(1)'}) = \mathrm{wt}(\mathbf{r}^{(2)'}) = \mathrm{wt}(\mathbf{e}') \in \{0, 2, \ldots, 2w_r\}$. This corresponds to the special case of $\mathbf{x} = \mathbf{x}'$ and $\mathbf{y} = \mathbf{y}'$ for the elements of $\Delta\mathcal{E}$ which are of the form

$$\mathbf{r}^{(2)} \cdot \mathbf{x} + \mathbf{r}^{(1)} \cdot \mathbf{y} + \mathbf{e} + \mathbf{r}^{(2)'} \cdot \mathbf{x}' + \mathbf{r}^{(1)'} \cdot \mathbf{y}' + \mathbf{e}'.$$

Theorem 1 (GV-like bound with error structure and $\mathbf{x}, \mathbf{y}$) *Let the decoder know $\mathbf{x}, \mathbf{y}$. Then, there exists a code $\mathscr{C} \subset \mathbb{F}_2^n$ of dimension*

$$k \geq n - 2w \log_2(\tfrac{n \cdot e}{w}) - 6w_r \log_2(\tfrac{n \cdot e}{2w_r}) - \log_2(w_r)$$

that can correct arbitrary HQC errors, i.e., guarantee correct decryption.

Proof. According to Lemma 5, we require $\mathscr{C} \cap \Delta\mathcal{E}^* = \{\mathbf{0}\}$. To bound the number of elements in $\Delta\mathcal{E}^*$, observe that

$$|\Delta\mathcal{E}^*| = \left| \bigcup_{\mathbf{x},\mathbf{y}:\mathrm{wt}(\mathbf{x})=\mathrm{wt}(\mathbf{y})=w} \Delta\mathcal{E}_{\mathbf{x},\mathbf{y}} \right| \leqslant \binom{n}{w}^2 |\Delta\mathcal{E}_{\mathbf{x},\mathbf{y}}| \leqslant \binom{n}{w}^2 \left(\sum_{i=0}^{2w_r} \binom{n}{i} \right)^3$$
$$\leqslant \left(\frac{n \cdot e}{w} \right)^{2w} w_r \left(\frac{n \cdot e}{2w_r} \right)^{6w_r},$$

where we have used $\binom{a}{b} \leqslant \left(\frac{a \cdot e}{b}\right)^b$. The statement follows due to Lemma 3. □

The approach to estimating the sizes of $\Delta\mathcal{E}$ and $\Delta\mathcal{E}^*$ in Lemma 4 and in Theorem 1 is imprecise in general since it ignores the possibility of collisions. Therefore, we expect that these bounds can be tightened.

Nevertheless, Theorem 1 already shows the potential of considering the error structure. This becomes clear from Table 1, which gives an overview of the bounds described in this section. The used error model, the implied decoding algorithm, and the resulting DFR are compared. Further, we compare the lower bounds on minimum required lengths under Heuristic 1 with the code lengths that are achievable by considering the error structure. For simplicity, w, w_r, and w_e were fixed to the choice of HQC in round 4. Note that higher n would require increasing these parameters, while a smaller code length might allow reducing them. Even when this effect is disregarded, it can already be observed that shorter code lengths are enabled by incorporating the error structure. This observation becomes more pronounced for higher security levels but is already significant for $\lambda = 128$. The following example elaborates on this finding.

Table 1. Overview of the bounds described in this section and comparison with the current HQC instantiation of round 4. Required and achievable code lengths for code dimension $k = \lambda$.

	HQC [9]	Lemma 1	Lemma 2	Lemma 4	Theorem 1
error model	BSC-like	BSC-like	BSC-like	structured	structured
decoder	multistage	ML	unique	unique	unique + $\mathbf{x}, \mathbf{y}$
DFR	$2^{-\lambda}$	$2^{-\lambda}$	0	0	0
	used n	LB on required n		UB on achievable n	
NIST 1	17669	13438	21882	5417	3800
NIST 3	35851	27913	49403	8243	5782
NIST 5	57637	45150	83767	10804	7576

Example 2. Again, we consider the HQC parameters of round 4 that achieve NIST-1 security, i.e., $\lambda \approx 128$ bit. That is, $n = 17669$, $w_r = 75$, $w = 66$ and the code concatenation has a dimension of $k = 128$. However, the current code-decoder design does not factor in the particular structure of the error vectors. In doing so, Lemma 4 guarantees that a code of length $n \leqslant 5417$ exists that can guarantee correct decoding. By taking $\mathbf{x}$ and $\mathbf{y}$ into account, Theorem 1 can reduce the upper bound on the required length to $n \leqslant 3800$.

The described improvement would increase the bandwidth efficiency of HQC considerably. Note, however, that the result is non-constructive: The proposed GV-like bounds do not provide an explicit code construction. Further, whether a computationally efficient decoder exists that accounts for the error structure remains open. We approach this second question in the following section.

4 Exploiting x and y to Reconstruct $\mathbf{r}^{(1)}$ and $\mathbf{r}^{(2)}$

In this section, we describe how, exploiting knowledge of $\mathbf{x}$ and $\mathbf{y}$, one can estimate coefficients of $\mathbf{r}^{(1)}$ and $\mathbf{r}^{(2)}$. We first present the general idea, then analyze it for the case of $\mathscr{C}_2$ being a repetition code because this makes the analysis easier. Indeed, in this case, we can get a rather tight theoretical estimate for the probability distribution of the number of correctly guessed coefficients.

4.1 Decoding as a Noisy Syndrome Decoding Problem

The idea is that, on input $\mathbf{c} = \mathbf{t} + \mathbf{z}$, with $\mathbf{t} \in \mathscr{C}$, one can use decoding through the inner code $\mathscr{C}_2$ to get an estimate of the error term $\mathbf{z}$. From this estimate, one can then guess coefficients of $\mathbf{r}^{(1)}$ and $\mathbf{r}^{(2)}$. These guesses are later used to remove some of the noise in $\mathbf{c}$. Indeed, let $\widehat{\mathbf{r}}^{(1)} \in \mathcal{R}$ and $\widehat{\mathbf{r}}^{(2)} \in \mathcal{R}$ denote the estimates for $\mathbf{r}^{(1)}$ and $\mathbf{r}^{(2)}$, respectively. Then, one can update $\mathbf{c}$ as

$$\begin{aligned}\widehat{\mathbf{c}} &= \mathbf{c} - \mathbf{x}\cdot\widehat{\mathbf{r}}^{(2)} - \mathbf{y}\cdot\widehat{\mathbf{r}}^{(1)}\\ &= \mathbf{t} + \mathbf{x}\cdot\widehat{\mathbf{r}}^{(2)} - \mathbf{y}\cdot\widehat{\mathbf{r}}^{(1)}\\ &= \mathbf{t} + \underbrace{\mathbf{x}\cdot\underbrace{\left(\mathbf{r}^{(2)} - \widehat{\mathbf{r}}^{(2)}\right)}_{\Delta\mathbf{r}^{(2)}} + \mathbf{y}\cdot\underbrace{\left(\mathbf{r}^{(1)} - \widehat{\mathbf{r}}^{(1)}\right)}_{\Delta\mathbf{r}^{(1)}} + \mathbf{e}}_{\mathbf{z}^*} = \mathbf{t} + \mathbf{z}^*.\end{aligned}$$

Notice that, if the estimates $\widehat{\mathbf{r}}^{(1)}$ and $\widehat{\mathbf{r}}^{(2)}$ are accurate, then $\Delta\mathbf{r}^{(1)}$ and $\Delta\mathbf{r}^{(2)}$ will have a low Hamming weight, perhaps considerably lower than w_r (remember that w_r is the weight of $\mathbf{r}^{(1)}$ and $\mathbf{r}^{(2)}$). If this happens, then the weight of $\mathbf{z}^*$ is with a high probability smaller than that of $\mathbf{z}$: Consequently, the decoding of $\widehat{\mathbf{c}}$ is more probable to succeed than the decoding of the original $\mathbf{c}$.

Estimating $\mathbf{r}^{(1)}$ and $\mathbf{r}^{(2)}$. Let us write $\mathbf{t} = \left(\mathbf{t}^{(1)}||\mathbf{t}^{(2)}||\cdots||\mathbf{t}^{(n/n_2)}\right)$ with each $\mathbf{t}^{(i)}$ being a codeword of $\mathscr{C}_2$, thus, of length n_2. Analogously, we write $\mathbf{z} = \left(\mathbf{z}^{(1)}||\mathbf{z}^{(2)}||\cdots||\mathbf{z}^{(n/n_2)}\right)$ where, again, each $\mathbf{z}^{(i)}$ has length n_2. Then, we can write $\mathbf{c} = \left(\mathbf{c}^{(1)}||\mathbf{c}^{(2)}||\cdots||\mathbf{c}^{(n/n_2)}\right)$, with $\mathbf{c}^{(i)} = \mathbf{t}^{(i)} + \mathbf{z}^{(i)}$.

Notice that since $\mathbf{z}$ has low weight, each $\mathbf{z}^{(i)}$ has low weight with high probability. Thus, with high probability decoding of $\mathbf{c}^{(i)}$ returns $\mathbf{t}^{(i)}$. Even when decoding fails, the output of the decoder is expected to be a codeword close to $\mathbf{t}^{(i)}$ (since the weight of $\mathbf{z}^{(i)}$ is generically low). We resume this reasoning by indicating $\widetilde{\mathbf{t}}^{(i)} = \mathsf{Dec}\left(\mathbf{c}^{(i)}\right)$ where, again, $\widetilde{\mathbf{t}}^{(i)} = \mathbf{t}^{(i)}$ holds with large probability. Then, we get a presumably good estimate of $\mathbf{z}$ by computing

$$\begin{aligned}\widehat{\mathbf{z}} &= \mathbf{c} - \left(\widetilde{\mathbf{t}}^{(1)}||\widetilde{\mathbf{t}}^{(2)}||\cdots||\widetilde{\mathbf{t}}^{(n/n_2)}\right)\\ &= \left(\mathbf{t}^{(1)}||\mathbf{t}^{(2)}||\cdots||\mathbf{t}^{(n/n_2)}\right) - \left(\widetilde{\mathbf{t}}^{(1)}||\widetilde{\mathbf{t}}^{(2)}||\cdots||\widetilde{\mathbf{t}}^{(n/n_2)}\right) + \mathbf{z} = \Delta\mathbf{t} + \mathbf{z}.\end{aligned}$$

Remember that $\mathbf{z} = \mathbf{x}\cdot\mathbf{r}^{(2)} + \mathbf{y}\cdot\mathbf{r}^{(1)} + \mathbf{e}$, hence

$$\widehat{\mathbf{z}} = \mathbf{x}\cdot\mathbf{r}^{(2)} + \mathbf{y}\cdot\mathbf{r}^{(1)} + \mathbf{e} + \Delta\mathbf{t}.$$

Since $\mathbf{e}$ has low weight by design and $\Delta\mathbf{t}$ is expected to have moderately low weight because of the above reasoning, the vector $\widehat{\mathbf{z}}$ can be seen as a noisy version of the vector $\left(\mathbf{x}\cdot\mathbf{r}^{(2)}||\mathbf{y}\cdot\mathbf{r}^{(1)}\right)$.

Let $\mathbf{r} \in \mathbb{F}_2^{2n}$ be the vector formed by the coefficients of $\mathbf{r}^{(2)}$ followed by those of $\mathbf{r}^{(1)}$. Moreover, let $\mathbf{H} = \left(\mathrm{Rot}(\mathbf{x})||\mathrm{Rot}(\mathbf{y})\right)$, where $\mathrm{Rot}(\mathbf{x})$ and $\mathrm{Rot}(\mathbf{y})$ are the circulant matrices whose first rows are $\mathbf{x}$ and $\mathbf{y}$, respectively. Then, we have

$$\widehat{\mathbf{z}} = \mathbf{H}\mathbf{r}^\top + \Delta\mathbf{z}.$$

The problem can be formulated as follows: on input a sparse parity-check matrix $\mathbf{H} \in \mathbb{F}_2^{n\times 2n}$ and a noisy syndrome $\widehat{\mathbf{z}}$, find a vector $\mathbf{r} \in \mathbb{F}_2^{2n}$ with weight $2w_r$ such that $\mathrm{wt}(\widehat{\mathbf{z}} - \mathbf{H}\mathbf{r}^\top)$ is minimum. This is precisely a Noisy Syndrome Decoding instance [6].

Input: noisy codeword $\mathbf{c} \in \mathbb{F}_2^n$, threshold $T \in \mathbb{N}$, secret key $(\mathbf{x}, \mathbf{y}) \in \mathcal{R}_w^2$, parity-check matrix $\mathbf{H} = \big(\mathrm{Rot}(\mathbf{x})||\mathrm{Rot}(\mathbf{y})\big) \in \mathbb{F}_2^{n\times 2n}$
Output: noisy codeword $\widehat{\mathbf{c}} \in \mathbb{F}_2^n$

/* Estimate error vector by decoding $\mathscr{C}_2$ */
1) compute $\mathbf{t} = \mathsf{Dec}_2(\mathbf{c})$
2) set $\widehat{\mathbf{z}} = \mathbf{c} - \mathbf{t}$

/* Guess coefficients of $\mathbf{r}^{(2)}$ */
3) Set $\widehat{\mathbf{r}}^{(2)} = (0, \cdots, 0)$;
4) for $i = 1, \cdots, n$, do:
5) set $b_i = \widehat{\mathbf{z}} \star \mathbf{h}_i$;// i-th counter
6) if $b_i \geqslant T$, set $\widehat{r}_i^{(2)} = 1$;

/* Guess coefficients of $\mathbf{r}^{(1)}$ */
7) Set $\widehat{\mathbf{r}}^{(1)} = (0, \cdots, 0)$
8) for $i = n+1, \cdots, 2n$, do:
9) set $b_i = \widehat{\mathbf{z}} \star \mathbf{h}_i$;// i-th counter
10) if $b_i \geqslant T$, set $\widehat{r}_i^{(1)} = 1$;

/* Update noisy codeword */
11) compute $\widehat{\mathbf{c}} = \mathbf{c} - \mathbf{x} \cdot \widehat{\mathbf{r}}^{(2)} - \mathbf{y} \cdot \widehat{\mathbf{r}}^{(1)}$.

Fig. 4. BF decoder to estimate coefficients of $\mathbf{r}^{(1)}$ and $\mathbf{r}^{(2)}$.

Any solver for this problem can then be used to retrieve $\mathbf{r}^{(1)}$ and $\mathbf{r}^{(2)}$. In the following, we consider a straightforward and efficient algorithm that allows us to guess a relatively large number of coefficients in $\mathbf{r}^{(1)}$ and $\mathbf{r}^{(2)}$.

Using Bit Flipping. Figure 4 describes the algorithm we consider. In the algorithm, we denote by $\mathbf{h}_i$ the i-th column of $\mathbf{H}$ and by $\star$ the integer product between two vectors defined over $\mathbb{F}_2$.

The algorithm is a translation of the basic BF decoder for LDPC codes. For every i, the algorithm computes the counter $\widehat{\mathbf{z}} \star \mathbf{h}_i$ and guesses a non-zero coefficient in $\mathbf{r}$ if the corresponding counter is high enough. Notice that the decoder output corresponds to the updated version of $\mathbf{c}$.

4.2 The Case of Repetition Codes

We show that, when the inner code is a repetition code, one can theoretically estimate the probability distribution of $\mathrm{wt}(\Delta\mathbf{r}^{(1)})$ and $\mathrm{wt}(\Delta\mathbf{r}^{(2)})$. In particular, we assume the distribution is identical for both polynomials. First, we need the following technical propositions, with their proofs in the appendix.

Proposition 1 (Probability to wrongly guess a 1-coefficient)
Let $i \in \{1,2\}$ and $j \in \{1, \cdots, n\}$ such that $r_j^{(i)} = 1$. Let the inner code be a

repetition code of odd length n_2 *and error-correction capability* $t = \lfloor \frac{n_2-1}{2} \rfloor$*. We denote the probability that* $\widehat{r}_j^{(i)} = 0$ *as* $\tau_{1\to 0}$ *and estimate*

$$\tau_{1\to 0} = \sum_{\ell=0}^{T-1} \binom{w}{\ell} \widehat{\rho}^{\,\ell} (1-\widehat{\rho})^{w-\ell},$$

with T *the BF threshold and*

$$\widehat{\rho} = (1-\widetilde{\rho}) \sum_{\ell=0}^{t-1} \binom{n_2-1}{\ell} \rho_z^\ell (1-\rho_z)^{n_2-1-\ell} + \widetilde{\rho} \sum_{\ell=t}^{n_2-1} \binom{n_2-1}{\ell} \rho_z^\ell (1-\rho_z)^{n_2-1-\ell},$$

where

$$\begin{aligned}\widetilde{\rho} = \widetilde{\rho}_{w,w_r-1} &\left(1 - \rho_{w,w_r}\left(1-\frac{w_e}{n}\right) - (1-\rho_{w,w_r})\frac{w_e}{n}\right) \\ &+ (1-\widetilde{\rho}_{w,w_r-1})\left(\rho_{w,w_r}\left(1-\frac{w_e}{n}\right) + (1-\rho_{w,w_r})\frac{w_e}{n}\right),\end{aligned}$$

$$\widetilde{\rho}_{w,w_r-1} = \sum_{\substack{\ell \in [1;\min\{w-1,w_r-1\}] \\ \ell \text{ odd}}} \frac{\binom{w-1}{\ell}\binom{n-w}{w_r-1-\ell}}{\binom{n-1}{w_r-1}}.$$

Proposition 2 (Probability to wrongly guess a 0-coefficient). *Let* $i \in \{1,2\}$ *and* $j \in \{1,\cdots,n\}$ *such that* $r_j^{(i)} = 0$*. Let the inner code be a repetition code of odd length* n *and error-correction capability* $t = \lfloor \frac{n-1}{2} \rfloor$*. We denote the probability that* $\widehat{r}_j^{(i)} = 1$ *as* $\tau_{0\to 1}$ *and estimate*

$$\tau_{0\to 1} = \sum_{\ell=T}^{w} \binom{w}{\ell} \widehat{\rho}^{\,\ell} (1-\widehat{\rho})^{w-\ell},$$

with T *the BF threshold and*

$$\widehat{\rho} = \widetilde{\rho} \sum_{\ell=0}^{t-1} \binom{n_2-1}{\ell} \rho_z^\ell (1-\rho_z)^{n_2-1-\ell} + (1-\widetilde{\rho}) \sum_{\ell=t}^{n_2-1} \binom{n_2-1}{\ell} \rho_z^\ell (1-\rho_z)^{n_2-1-\ell},$$

where

$$\begin{aligned}\widetilde{\rho} = \widetilde{\rho}_{w,w_r} &\left(1 - \rho_{w,w_r}\left(1-\frac{w_e}{n}\right) - (1-\rho_{w,w_r})\frac{w_e}{n}\right) \\ &+ (1-\widetilde{\rho}_{w,w_r})\left(\rho_{w,w_r}\left(1-\frac{w_e}{n}\right) + (1-\rho_{w,w_r})\frac{w_e}{n}\right),\end{aligned}$$

$$\widetilde{\rho}_{w,w_r} = \sum_{\substack{\ell \in [1;\min\{w-1,w_r\}] \\ \ell \text{odd}}} \frac{\binom{w-1}{\ell}\binom{n-w}{w_r-\ell}}{\binom{n-1}{w_r}}.$$

Using the above results, we can derive the weight distribution for each $\Delta\mathbf{r}^{(i)}$.

Proposition 3 (Weight distribution of $\Delta\mathbf{r}^{(i)}$). *The probability that $\Delta\mathbf{r}^{(i)}$ has weight $w_r^{(i)}$ can be estimated as*

$$\Pr\left[\mathrm{wt}(\Delta\mathbf{r}^{(i)}) = w_r^{(i)}\right] = \sum_{j=0}^{\min\{w_r, w_r^{(i)}\}} \Pr\left[N_1 = j\right] \Pr\left[N_0 = w_r^{(i)} - j\right],$$

where N_0 and N_1 denote number of wrongly guessed 0-coefficients and 1-coefficients, which follow the distribution

$$\Pr\left[N_0 = j\right] = \binom{n - w_r}{j} (\tau_{0\to 1})^j (1 - \tau_{0\to 1})^{n - w_r - j},$$

$$\Pr\left[N_1 = j\right] = \binom{w_r}{j} (\tau_{1\to 0})^j (1 - \tau_{1\to 0})^{w_r - j}.$$

Proof. Whenever a one-coefficient of $\mathbf{r}^{(2)}$ is guessed correctly, the weight of $\Delta\mathbf{r}^{(2)}$ decreases by 1, while a wrong estimate does not change the weight of $\Delta\mathbf{r}^{(2)}$ in comparison with $\mathbf{r}^{(2)}$. Analogously, guessing a zero-coefficient correctly does not change the weight, while a wrong guess increases the weight by 1. Let N_0 denote the number of wrongly guessed 0-coefficients and N_1 the number of wrongly guessed 1-coefficients. Then, the weight of $\Delta\mathbf{r}^{(2)}$ is $N_0 + N_1$, and

$$\Pr\left[\mathrm{wt}(\Delta\mathbf{r}^{(i)}) = w_r^{(i)}\right] = \sum_{j=0}^{\min\{w_r, w_r^{(i)}\}} \Pr\left[N_1 = j, N_0 = w_r^{(i)} - j\right].$$

To conclude the proof, it is enough to assume that coefficients are guessed independently so that both N_0 and N_1 are the sums of Bernoulli variables with parameters $\tau_{0\to 1}$ and $\tau_{1\to 0}$.

Empirical Validation. The above formulas are somewhat convoluted: many parameters are employed, and evaluating their interplay is difficult. Further, Proposition 3 models the coefficients of $\Delta\mathbf{r}^{(i)}$ as independent, which is not valid in general, as the inner decoding operates on multiple symbols of $\mathbf{c}$ jointly. We can comment on the effectiveness of the proposed approach using numerical simulations; to this end, consider Fig. 5.

As we can see, the approximation as independent random variables works well in practice. Further, there are values of T for which both $\tau_{1\to 0}$ and $\tau_{0\to 1}$ are low. That is, the probability of getting a wrong estimate for a coefficient is low, regardless of its value. This guarantees that both $\widehat{\mathbf{r}}^{(1)}$ and $\widehat{\mathbf{r}}^{(2)}$ are close to $\mathbf{r}^{(1)}$ and $\mathbf{r}^{(2)}$, which, in turn, implies that the weights of $\Delta\mathbf{r}^{(1)}$ and $\Delta\mathbf{r}^{(2)}$ are decreased with high probability.

Notice also that if T is too low, then too many 0 s are estimated as 1 s, and the weight of $\Delta\mathbf{r}^{(1)}$ and $\Delta\mathbf{r}^{(2)}$ may even be larger than that of $\mathbf{r}^{(1)}$ and $\mathbf{r}^{(2)}$.

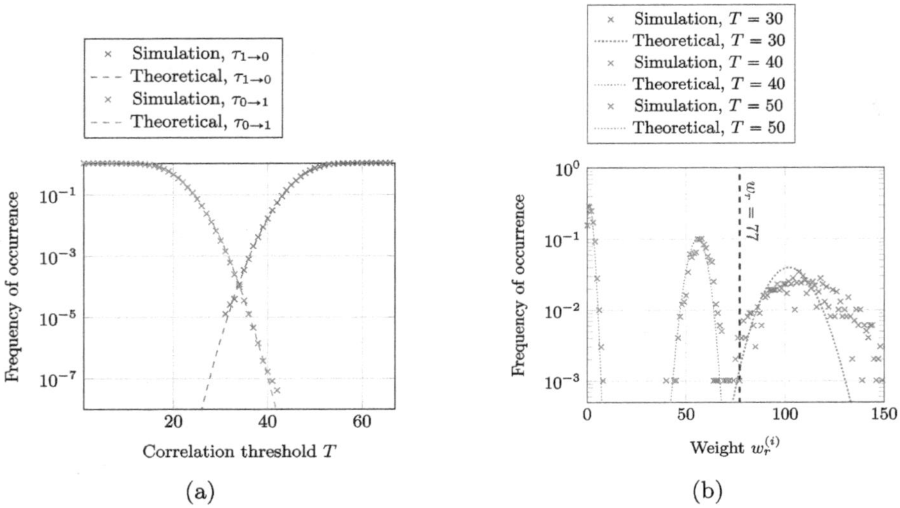

Fig. 5. Comparison of numerical and theoretical estimates for HQC version BCH⊗REP. Round-1 parameters of HQC offering 128 bits of security were used, i.e., $n_1 = 766$, $n_2 = 31$, $w = 67$, and $w_r = w_e = 77$. Figure 5a shows the values of $\tau_{1\to0}$ and $\tau_{0\to1}$, Figure 5b reports the distribution of the weight of $\Delta\mathbf{r}^{(1)}$ and $\Delta\mathbf{r}^{(2)}$. The numerical estimates have been obtained averaging over 10^3 trials.

5 Reducing the DFR with a Structure-Aware Decoder

We now show that the technique from the previous section improves HQC decoding: for the same code length n, our decoder has a lower DFR than the original HQC decoder.

The algorithm, summarized in Fig. 6, can be viewed as a conventional HQC decoder, enhanced by an initial noise-reducing stage performed according to the algorithm in Fig. 4. The final decoding step requires decoding the inner codes a second time, which gives the proposed algorithm an iterative nature. The results of the second inner decoding correlate with the results of the initial decoding: inner blocks that failed in the initial phase are more likely to fail again in the final decoding step. As a consequence of these dependencies, we were not able to derive a reliable DFR model. While the analysis of the proposed decoder is complex, its running time remains efficient, as the following proposition demonstrates.

Proposition 4. (Complexity of proposed decoder). *The computational complexity of the proposed decoder is in*

$$\mathcal{O}\left(T_{\text{inner}} + T_{\text{outer}} + nw + ww_r\right),$$

where T_{inner} *and* T_{outer} *denote the running times of the inner and outer decoders.*

Input: noisy codeword $\mathbf{c} \in \mathbb{F}_2^n$, secret key $\mathbf{x}, \mathbf{y}$
Output: decrypted message $\mathbf{m}$

```
/* Estimate error vector by decoding 𝒞₂ */
```
1) set $\widehat{\mathbf{z}} = \mathbf{c} - \mathsf{Dec}_2(\mathbf{c})$;

```
/* Estimate r̂(1), r̂(2) */
```
2) derive $\widehat{\mathbf{r}}^{(1)}, \widehat{\mathbf{r}}^{(2)}$ as in Figure 4;

```
/* Update noisy codeword */
```
3) compute $\widehat{\mathbf{c}} = \mathbf{c} - \mathbf{x} \cdot \widehat{\mathbf{r}}^{(2)} - \mathbf{y} \cdot \widehat{\mathbf{r}}^{(1)}$;

```
/* standard HQC decoder */
```
4) decode $\mathbf{m} = \mathsf{Dec}(\widehat{\mathbf{c}})$.

Fig. 6. Proposed error structure-aware decoder for HQC.

Proof. The initial inner decoding has cost T_{inner}. In the noisy syndrome decoding step, each counter is computed using $\mathcal{O}(w)$ operations since w coefficients are summed. Therefore, the estimation of $\widehat{\mathbf{r}}^{(1)}, \widehat{\mathbf{r}}^{(2)}$ takes time $\mathcal{O}(nw)$. For a properly chosen threshold T, the weights of the polynomials $\widehat{\mathbf{r}}^{(i)}$ do not exceed w_r. Hence, $\mathbf{x}\widehat{\mathbf{r}}^{(2)} + \mathbf{y}\widehat{\mathbf{r}}^{(1)}$ requires $\mathcal{O}(ww_r)$ operations leveraging the sparse nature of the polynomials. The final HQC decoding procedure has cost $T_{\text{inner}} + T_{\text{outer}}$. □

Remark 1. The proposition considers a non-constant-time implementation of the decoder. Still, we expect that a constant time implementation does not lead to a significant penalty: indeed, any of the strategies employed for BF decoders can probably be employed for our decoder as well.

Proposition 4 shows that the increase in complexity due to the additional noise filtering step is rather limited due to the sparsity of the involved polynomials. Nevertheless, a considerable improvement in decoding performance can be achieved as can be observed empirically, see Fig. 7.

In Fig. 7a, we focus on the concatenation of BCH and repetition code. The DFR of the decoder is simulated for several threshold values T while varying the inner code length n_2. It can be observed that the optimal threshold value for this setup is $T = 38$, which results in a DFR that is 10^9 times lower than the one of the original decoder when $n_2 = 21$.

In Fig. 7b, the concatenation of Reed-Muller and Reed-Solomon code is considered. Since the length of the inner RM code is not flexible, we vary the outer length n_1 instead. Here, the optimal threshold is identified as $T = 39$. Although the improvement is not as substantial as for the concatenation of BCH and repetition code, it is still evident that the DFR of the proposed decoding algorithm is significantly lower than that of the original decoder.

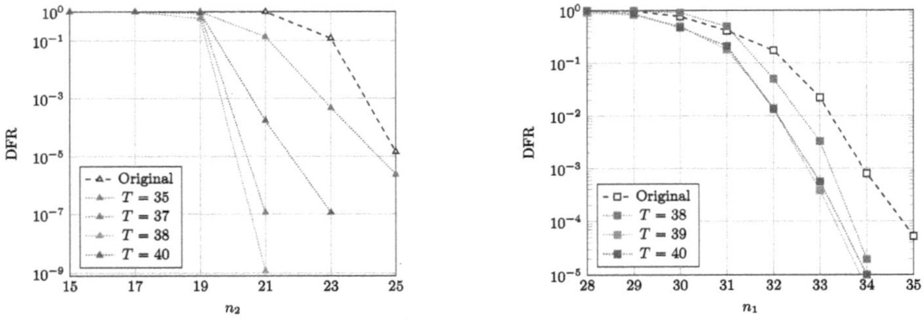

(a) BCH ⊗ REP (b) RS ⊗ DRM

Fig. 7. Simulation of the DFR of the original HQC decoder and the proposed one, as a function of either n_1 or n_2. Figure 7a considers the BCH ⊗ REP version with $n_1 = 766$, $t_{\mathrm{BCH}} = 57$, $w = 67$, $w_r = 77$; Fig. 7b considers the RS ⊗ DRM version with $n_{\mathrm{RM}} = 128$ (original length RM code), $mult = 3$ (multiplication factor RM code), $w = 66$, $w_r = w_e = 75$.

6 Conclusion

In this work, we have taken an initial step towards replacing the BSC model currently used in HQC with a channel model that accounts for the specific error structure tied to the secret key.

First, we analyzed the potential improvements in code length that can be achieved. The derived GV-like bounds show that significantly shorter codes than those currently used in HQC can guarantee correct decoding. This stands in contrast to the limited room for improvements under the BSC model. However, while our study reveals a promising direction for more effective code-decoder pairs, it is non-constructive.

Second, we proposed a new error structure-aware decoder. This decoder is a modification of the standard multistage decoder of HQC, incorporating an additional filtering stage. The filtering step leverages the secret key to estimate coefficients of the unknown polynomials that form the error vector. As a result, the weight of errors to be corrected can be reduced, potentially leading to a lower DFR. We have developed a preliminary analysis of the behavior of the proposed decoder. However, precisely modeling the DFR remains a challenge due to the iterative nature of the algorithm.

Acknowledgements. Marco Baldi is supported by the Italian Ministry of University and Research (MUR) PRIN 2022 program under the "Mathematical Primitives for Post Quantum Digital Signatures" (P2022J4HRR) and "Post quantum Identification and eNcryption primiTives: dEsign and Realization (POINTER)" (2022M2JLF2) projects, and under the Italian Fund for Applied Science (FISA 2022), Call for tender No. 1405 published on 13-09-2022 by MUR - project title "Quantum-safe cryptographic tools for the protection of national data and information technology assets" (QSAFEIT) -

No. FISA 2022-00618 - CUP I33C24000520001 - Grant Assignment Decree no. 15461 adopted on 02.08.2024 by the Italian MUR.

Sebastian Bitzer acknowledges the financial support by the Federal Ministry of Education and Research of Germany in the program of "Souverän. Digital. Vernetzt.". Joint project 6G-life, project identification number: 16KISK002.

A Proofs of Propositions 1 and 2

We require two preliminary results, which we give and prove here.

Proposition 5. *For two polynomials* $\mathbf{a} \in \mathcal{R}_{w_a}$ *and* $\mathbf{b} \xleftarrow{\$} \mathcal{R}_{w_b}$, *an arbitrary coefficient (say, the first one) in the product* $\mathbf{a} \cdot \mathbf{b}$ *is set with probability*

$$\rho_{w_a, w_b} = \sum_{\substack{\ell \in [1; \min\{w_a, w_b\}] \\ \ell \text{ odd}}} \frac{\binom{w_a}{\ell}\binom{n-w_a}{w_b-\ell}}{\binom{n}{w_b}}.$$

Proof. According to the rules of polynomial multiplication, we have:

$$c_\ell = \sum_{\substack{i,j \\ i+j \equiv \ell \mod n}} a_i \cdot b_j, \quad \text{for } \ell \in \{0, 1, \dots, n-1\}.$$

We have $c_\ell = 1$ only if, in the above equation, the number of terms $a_i \cdot b_j$ equal to 1 is odd. The probability to have exactly ℓ terms $a_i \cdot b_j$ which are equal to 1 is $\frac{\binom{w_a}{\ell}\binom{n-w_a}{w_b-\ell}}{\binom{n}{w_b}}$. □

Proposition 6. *Let* $\mathbf{x}, \mathbf{y} \in \mathcal{R}_w$, $\mathbf{r}^{(1)} \xleftarrow{\$} \mathcal{R}_{w_r}$, $\mathbf{r}^{(2)} \xleftarrow{\$} \mathcal{R}_{w_r}$, *and* $\mathbf{e} \xleftarrow{\$} \mathcal{R}_{w_e}$. *An arbitrary coefficient of the polynomial* $\mathbf{z} = \mathbf{x} \cdot \mathbf{r}^{(2)} + \mathbf{y} \cdot \mathbf{r}^{(1)} + \mathbf{e}$ *follows a Bernoulli distribution with parameter*

$$\rho_z = 4\rho_{w,w_r}^2 \rho_e - 2\rho_{w,w_r}^2 - 4\rho_{w,w_r}\rho_e + 2\rho_{w,w_r} + \rho_e.$$

Under the assumption that the coefficients behave as independent random variables, the weight of $\mathbf{z}$ *follows a binomial distribution.*

Proof. According to Proposition 5, the i-th coefficient of both products $\widehat{\mathbf{x}} = \mathbf{x} \cdot \mathbf{r}^{(2)}$ and $\widehat{\mathbf{y}} = \mathbf{y} \cdot \mathbf{r}^{(1)}$ is Bernoulli distributed with parameter $\rho_{w,w_r^{(2)}}$ and $\rho_{w,w_r^{(1)}}$. Then, the i-th coefficient of $\mathbf{z}$ will be set with probability

$$\begin{aligned} &\Pr[\widehat{x}_i = 1] \cdot \Pr[\widehat{y}_i = 0] \cdot \Pr[e_i = 0] + \Pr[\widehat{x}_i = 0] \cdot \Pr[\widehat{y}_i = 1] \cdot \Pr[e_i = 0] \\ + &\Pr[\widehat{x}_i = 0] \cdot \Pr[\widehat{y}_i = 0] \cdot \Pr[e_i = 1] + \Pr[\widehat{x}_i = 1] \cdot \Pr[\widehat{y}_i = 1] \cdot \Pr[e_i = 1]. \end{aligned}$$

Substituting the probabilities with the associated Bernoulli parameters, we get

$$\begin{aligned} \rho_z = &(1 - \rho_{w,w_r})(1 - \rho_{w,w_r})\rho_e + (1 - \rho_{w,w_r})\rho_{w,w_r}(1 - \rho_e) \\ &+ \rho_{w,w_r}(1 - \rho_{w,w_r})(1 - \rho_e) + \rho_{w,w_r}\rho_{w,w_r}\rho_e. \end{aligned}$$

After some manipulations, we obtain the expression for ρ_z. □

We now proceed to prove Proposition 1; the proof for Proposition 2 is carried out in the same way and is hence omitted.

A.1 Proof of Proposition 1

For $i \in \text{supp}(\mathbf{r}^{(2)})$, we derive the probability distribution of the counter value. We express $\mathbf{r}^{(2)}$ as

$$\mathbf{r}^{(2)} = \underbrace{(0, \cdots, 0, 1, 0, \cdots, 0)}_{\substack{1 \text{ in position } j, \\ 0 \text{ elsewhere}}} + \widetilde{\mathbf{r}}^{(2)},$$

where $\widetilde{\mathbf{r}}^{(2)}$ is equal to $\mathbf{r}^{(2)}$ apart from the coefficient in position j, which is 0 (instead of 1). Observe that $\text{wt}(\widetilde{\mathbf{r}}^{(2)}) = w_r - 1$. The counter associated to $r_j^{(2)}$ is obtained by summing the coefficients of the estimated error vector $\widehat{\mathbf{z}}$ in the positions indexed by $\text{supp}(\mathbf{x}) + j = \{s + j \mod n \mid s \in \text{supp}(\mathbf{x})\}$. For each $\ell \in \text{supp}(\mathbf{x}) + j$, we have:

$$z_\ell = 1 + \underbrace{(\mathbf{x} \cdot \widetilde{\mathbf{r}}^{(2)})_\ell + (\mathbf{y} \cdot \mathbf{r}^{(1)} + \mathbf{e})_\ell}_{\widetilde{z}_\ell} = 1 + \widetilde{z}_\ell,$$

where z_ℓ is the ℓ-th coefficient of $\widetilde{\mathbf{z}} = \mathbf{x} \cdot \widetilde{\mathbf{r}}^{(2)} + \mathbf{y} \cdot \mathbf{r}^{(1)} + \mathbf{e}$. With arguments analogous to those in Proposition 6, we see that $(\mathbf{y} \cdot \mathbf{r}^{(1)} + \mathbf{e})_j$ is Bernoulli distributed with parameter

$$\rho_{w,w_r}\left(1 - \frac{w_e}{n}\right) + (1 - \rho_{w,w_r})\frac{w_e}{n}.$$

Also $(\mathbf{x} \cdot \widetilde{\mathbf{r}}^{(2)})_j$ can be considered as Bernoulli distributed; in particular, the probability that its ℓ-th coefficient is 1 corresponds to

$$\widetilde{\rho}_{w,w_r-1} = \sum_{\substack{\ell \in [1;\min\{w-1,w_r-1\}] \\ \ell \text{ odd}}} \frac{\binom{w-1}{\ell}\binom{n-1-w}{w_r-1-\ell}}{\binom{n-1}{w_r-1}}.$$

Notice that this probability slightly differs from the one in Proposition 5. This is because we are considering only $w - 1$ coefficients of $\mathbf{x}$ and $n - 1$ coordinates for $\widetilde{\mathbf{r}}^{(2)}$ (as one is set to 0).

Putting everything together, we get that $\widetilde{z}_j$ is equal to 1 with probability

$$\begin{aligned}\widetilde{\rho} = \underbrace{\widetilde{\rho}_{w,w_r-1}}_{\Pr\left[(\mathbf{x}\cdot\widetilde{\mathbf{r}}^{(2)})_\ell=1\right]} & \underbrace{\left(1 - \rho_{w,w_r}\left(1 - \frac{w_e}{n}\right) - (1 - \rho_{w,w_r})\frac{w_e}{n}\right)}_{\Pr\left[(\mathbf{y}\cdot\mathbf{r}^{(1)}+\mathbf{e})_\ell=0\right]} \\ & + \underbrace{(1 - \widetilde{\rho}_{w,w_r-1})}_{\Pr\left[(\mathbf{x}\cdot\mathbf{r}^{(2)})_\ell=0\right]} \underbrace{\left(\rho_{w,w_r}\left(1 - \frac{w_e}{n}\right) + (1 - \rho_{w,w_r})\frac{w_e}{n}\right)}_{\Pr\left[(\mathbf{y}\cdot\mathbf{r}^{(1)}+\mathbf{e})_\ell=1\right]}\end{aligned}$$

Let $\perp_{\text{in}}(\ell)$ indicate the event that the inner code codeword in which the ℓ-th coordinate is contained is wrongly decoded. The complementary event (i.e., a decoding success) is indicated as $\bar{\perp}_{\text{in}}(\ell)$. Then, we have

$$\Pr\left[\widehat{z}_\ell = 1\right] = \Pr\left[\bar{\perp}_{\text{in}}(\ell) \mid \widetilde{z}_\ell = 0\right] \cdot \Pr\left[\widetilde{z}_\ell = 0\right] + \Pr\left[\perp_{\text{in}}(\ell) \mid \widetilde{z}_\ell = 1\right] \Pr\left[\widetilde{z}_\ell = 1\right]. \quad (3)$$

Indeed, the first term (i.e., $\Pr\left[\bar{\bot}_{\text{in}}(\ell) \mid \widetilde{z}_\ell = 0\right] \cdot \Pr\left[\widetilde{z}_\ell = 0\right]$) is the probability that $z_\ell = 1$ and decoding is successful, so z_ℓ is correctly estimated. Analogously, the second term (i.e., $\Pr\left[\bot_{\text{in}}(\ell) \mid \widetilde{z}_\ell = 1\right] \Pr\left[\widetilde{z}_\ell = 1\right]$) corresponds to the probability that $z_\ell = 0$ but there is a decoding failure, so z_ℓ is wrongly estimated.

Let $\text{supp}_{\text{in}}(\ell)$ denote the set of indices that correspond to the same inner codeword as position ℓ. Then, $\widehat{z}_\ell = z_\ell = 1$ if and only if the remaining $n_2 - 1$ positions allow correct decoding of position ℓ, despite position ℓ being erroneous. This happens whenever the number of set coefficients in $\text{supp}_{\text{in}}(\ell) \setminus \{\ell\}$ is not greater than $t - 1$, where $t = \lfloor \frac{n_2 - 1}{2} \rfloor$ denotes the error-correction capability of the inner repetition code. The DFR analysis of HQC assumes independence between the coefficients of the error $\mathbf{z}$ [3]. Similarly, we assume that the positions in $\text{supp}_{\text{in}}(\ell) \setminus \{\ell\}$ are independently Bernoulli distributed with parameter ρ_z as in Proposition 6. Then, the required probabilities can be calculated as

$$\Pr\left[\bar{\bot}_{\text{in}}(\ell) \mid \widetilde{z}_\ell = 0\right] = \sum_{k=0}^{t-1} \binom{n_2 - 1}{k} \rho_z^k (1 - \rho_z)^{n_2 - 1 - k}.$$

With analogous reasoning, we obtain

$$\Pr\left[\bot_{\text{in}}(j) \mid \widetilde{z}_j = 1\right] = \sum_{k=t}^{n_2 - 1} \binom{n_2 - 1}{k} \rho_z^k (1 - \rho_z)^{n_i - 1 - k}.$$

Then, writing $\widehat{\rho} = \Pr\left[\widehat{z}_\ell = 1\right]$, we obtain (3) as

$$\widehat{\rho} = (1 - \widehat{\rho}) \sum_{k=0}^{t-1} \binom{n_i - 1}{k} \rho_z^k (1 - \rho_z)^{n_2 - 1 - k} + \widetilde{\rho} \sum_{k=t}^{n_2 - 1} \binom{n_2 - 1}{k} \rho_z^k (1 - \rho_z)^{n_2 - 1 - k}. \quad (4)$$

We model $\sum_{j \in \text{supp}\mathbf{x}+i} \widehat{z}_j$ as a sum of independent random variables. Consequently, $\sum_{j \in \text{supp}\mathbf{x}+i} \widehat{z}_j$ follows the binomial distribution with w trials and success probability $\widehat{\rho}$ and we obtain

$$\tau_{1 \to 0} = \sum_{\ell=0}^{T-1} \binom{w}{\ell} \widehat{\rho}^{\,\ell} (1 - \widehat{\rho})^{w - \ell}.$$

References

1. C. Aguilar-Melchor et al. “Efficient error-correcting codes for the HQC post-quantum cryptosystem”. In: Designs, Codes and Cryptography (2024), pp. 1–20
2. Albrecht, M.R., et al.: Classic McEliece: conservative code-based cryptography (2020). https://classic.mceliece.org/nist/mceliece-20201010.pdf
3. Aragon, N., Gaborit, G., Zémor, P.: HQC-RMRS, an instantiation of the HQC encryption framework with a more efficient auxiliary error-correcting code. arXiv preprint arXiv:2005.10741 (2020)
4. Aragon, N., et al.: BIKE - bit flipping key encapsulation (2023). https://bikesuite.org/

5. Baldi, M., Barenghi, A., Chiaraluce, F., Pelosi, G., Santini, P.: LEDAcrypt: QC-LDPC code-based cryptosystems with bounded decryption failure rate. In: Baldi, M., Persichetti, E., Santini, P. (eds.) CBC 2019. LNCS, vol. 11666, pp. 11–43. Springer, Cham (2019). https://doi.org/10.1007/978-3-030-25922-8_2
6. Deneuville, J.-C., Gaborit, P., Zémor, G.: Ouroboros: a simple, secure and efficient key exchange protocol based on coding theory. In: Lange, T., Takagi, T. (eds.) PQCrypto 2017. LNCS, vol. 10346, pp. 18–34. Springer, Cham (2017). https://doi.org/10.1007/978-3-319-59879-6_2
7. Elias, P.: Coding for noisy channels. In: IRE WESCON Convention Record, vol. 2, pp. 94–104 (1955)
8. Loeliger, H.-A.: On the basic averaging arguments for linear codes. In: Communications and Cryptography: Two Sides of One Tapestry, pp. 251–261 (1994)
9. Melchor, C.A., et al.: Hamming quasi-cyclic (HQC). In: NIST PQC Round 2.4, p. 13 (2018)
10. NIST Computer Security Resource Center: Post-quantum cryptography standardization (2017). https://csrc.nist.gov/Projects/post-quantumcryptography/

Generalizing the Augot-Finiasz PKE to Other Code Classes

Anmoal Porwal[1(✉)], Anna Baumeister[1], Violetta Weger[1], Antonia Wachter-Zeh[1], and Pierre Loidreau[2]

[1] Technical University of Munich, Munich, Germany
{anmoal.porwal,anna.baumeister,violetta.weger,antonia.wachter-zeh}@tum.de
[2] DGA and Univ Rennes, CNRS, IRMAR - UMR 6625, F-35000 Rennes, France
pierre.loidreau@univ-rennes.fr

Abstract. The Augot-Finiasz system is a public-key encryption (PKE) scheme based on Reed-Solomon codes and was later followed by analogous versions in the rank metric. Although these schemes were eventually broken, their fundamental idea remains exciting. Notably, these schemes are significantly different from the McEliece system as there is no need to hide the code and, as such, promise much better parameters. Further, they admit a simple description where both the public key and ciphertext are just corrupted codewords of a public code. An interesting question is whether the general idea can be made to work, i.e., resist all known attacks, by using other code classes. This paper shows how to generalize the Augot-Finiasz system to other code families. We reduce the correctness and security of this framework to simple assertions about the code class with which it is instantiated. Specifically, its correctness is equivalent to the existence of an efficient error-erasure decoder, and its security reduces to an easily understood hardness assumption, called *supercode decoding*, close to the syndrome decoding problem. We provide a negative answer for various code families by showing that solving the supercode decoding problem is easy for them. It remains an open question whether a secure choice exists.

Keywords: public-key encryption · error-erasure decoding · code-based cryptography

1 Introduction

The McEliece public-key encryption (PKE) scheme is considered the leading code-based candidate in the NIST competition for standardizing post-quantum cryptography. However, it suffers from very large key sizes and relies on the so-called Goppa distinguishing assumption, which has recently come under increasingly severe attacks [16]. Hence, various alternatives have been proposed in the last decades.

V. Weger et al. (Eds.): CBCrypto 2024, LNCS 15531, pp. 25–44, 2025.
https://doi.org/10.1007/978-3-031-90229-1_2

The line of cryptosystems we will be interested in began their history in 2003 with the Augot and Finiasz [2,3] cryptosystem. They described a PKE using Reed-Solomon (RS) codes, where the security relies on a problem known as polynomial reconstruction. In contrast to the McEliece system, the scheme does not rely on hiding a code and it has a simple description where the public key and ciphertext could be seen as corrupted codewords of an RS code. Although the scheme required large field sizes, it provided much smaller keys than McEliece. In 2006, Faure and Loidreau [7] proposed a rank-metric version using Gabidulin codes, where the security was related to a linearized polynomial version of the polynomial reconstruction problem. Since generic decoding attacks in the rank metric are known to be much more expensive than in the Hamming metric, their proposed parameters were even smaller than those of Augot-Finiasz. However, the Hamming and rank-metric schemes were eventually broken by Coron [6] and Gaborit, Otmani, and Talé–Kalachi [9], respectively. The history of these schemes concludes with two recent repairs of the Faure-Loidreau system: RAMESSES [12] and LIGA [17]. Unfortunately, these attempts did not succeed, as was shown by Bombar and Couvreur [5].

In the McEliece scheme, the ciphertext is a codeword plus an error, and the secret is some hidden property of the code which allows for efficient decoding. In this scheme, the ciphertext is a codeword plus an error, but the secret is some property *of the error vector* rather than the code. Hence, the code itself can be made public. Even though the scheme was broken when used with RS or Gabidulin codes, the general idea of this scheme is promising and somewhat natural. Therefore, an important question is whether this idea can be realized using a different code family.

1.1 Contributions

We introduce a PKE framework that generalizes the Augot-Finiasz and the Faure-Loidreau system to new code families. The correctness and security of this framework are equivalent to simple assertions about this family. Specifically, security reduces to a hard problem called *supercode decoding*, and correctness is equivalent to the existence of an efficient error-erasure decoder for this code class. We give a largely unified description of the Augot-Finiasz scheme and its successors, along with how they were attacked. Further, our reduction allows us to show that various other code families would be insecure as well. A question left open is whether there exists a secure choice for the code family or, conversely, if the framework (as stated) is uninstantiable. Even so, we think the general idea of this framework is promising and it is a somewhat natural counterpart to the McEliece paradigm. Additionally, while the framework as stated might be insecure, various modifications are open for further study (for example, using a different "randomization" of the public key in the encryption step), that might eventually lead to secure instantiations.

1.2 Organization of the Paper

Section 2 provides the required background and notation. We describe the framework in Sect. 3 and state the conditions for its correctness and security in Sect. 4. The previously proposed schemes this framework was inspired from are surveyed in Sect. 5. Finally, Sect. 6 describes which code classes are insecure and concludes the paper.

2 Preliminaries

Notation. The finite field of size q^m is denoted by $\mathbb{F}_{q^m}$ where q is a prime power and m is a nonzero integer. The notation $\mathbb{F}_{q^m}^*$ refers to the set of its nonzero elements. We write $\alpha \xleftarrow{\$} \mathcal{S}$ to mean that α is drawn uniformly at random from the set $\mathcal{S}$.

For a matrix $\mathbf{X} \in \mathbb{F}_{q^m}^{k \times n}$ and a vector $\boldsymbol{x} \in \mathbb{F}_{q^m}^n$, the rank of $\mathbf{X}$ over $\mathbb{F}_{q^m}$ is denoted by $\text{rank}(\mathbf{X})$ and the rank of the $\mathbb{F}_q$-space spanned by the entries $x_1, \dots, x_n$ of $\boldsymbol{x}$ is written $\text{rank}(\mathbf{x})$. For clarity, we will write $\text{rank}_{\mathbb{F}_{q^m}}(\mathbf{X})$ and $\text{rank}_{\mathbb{F}_q}(\boldsymbol{x})$. We denote by $\langle \mathbf{X} \rangle$ the $\mathbb{F}_{q^m}$-subspace spanned by the rows of $\mathbf{X}$ and by $\langle \boldsymbol{x} \rangle$ the $\mathbb{F}_q$-space spanned by the entries of $\boldsymbol{x}$. To avoid ambiguity, these are sometimes written as $\langle \mathbf{X} \rangle_{\mathbb{F}_{q^m}}$ and $\langle \boldsymbol{x} \rangle_{\mathbb{F}_q}$. Likewise, for a list of vectors $\boldsymbol{x}^{(1)}, \dots, \boldsymbol{x}^{(k)} \in \mathbb{F}_{q^m}^n$, their $\mathbb{F}_{q^m}$-span is $\langle \boldsymbol{x}^{(1)}, \dots, \boldsymbol{x}^{(k)} \rangle_{\mathbb{F}_{q^m}}$. If $\mathcal{U}$ is a subspace over $\mathbb{F}_q$, then $|\mathcal{U}|$ refers to its dimension, and if $\mathcal{U}$ is just a set, then the same notation denotes its cardinality.

We denote by $\mathcal{P}$, respectively $\mathcal{L}$, the set of polynomials, respectively linearized polynomials [14], in the variable x and coefficients in $\mathbb{F}_{q^m}$. Further, notation like $\mathcal{P}_{<k}$ ($\mathcal{L}_{<k}$) will denote (linearized) polynomials with (q-)degree less than k. For any (q-)polynomial f and a vector $\boldsymbol{x} \in \mathbb{F}_{q^m}^n$, we denote by $f(\boldsymbol{x})$ its element-wise evaluation, i.e., $(f(x_1), \dots, f(x_n)) \in \mathbb{F}_{q^m}^n$.

Finally, $\boldsymbol{x}^q$ denotes component-wise exponentiation, i.e., $(x_1^q, \dots, x_n^q) \in \mathbb{F}_{q^m}^n$ and, if $\boldsymbol{y} \in \mathbb{F}_{q^m}^n$, then $\boldsymbol{x} \cdot \boldsymbol{y} = (x_1 y_1, \dots, x_n y_n)$ is the component-wise product of the two vectors.

Hamming Metric. The Hamming weight of a vector in $\mathbb{F}_{q^m}^n$ is the number of its nonzero entries. The *support* of a vector $\boldsymbol{x} \in \mathbb{F}_{q^m}^n$ is denoted by $\text{supp}(\boldsymbol{x})$ and is the set of indices of its nonzero entries. In other words,

$$\text{supp}(\boldsymbol{x}) = \{i \in \{1, \dots, n\} \,|\, x_i \neq 0\}.$$

Rank Metric. The rank weight of a vector $\boldsymbol{x} \in \mathbb{F}_{q^m}^n$ over $\mathbb{F}_q$ is given by the dimension of the $\mathbb{F}_q$-subspace spanned by its entries. Let $\Gamma = \{\gamma_1, \dots, \gamma_m\}$ be a basis of $\mathbb{F}_{q^m}$ over $\mathbb{F}_q$. Then we can define the expansion map

$$\begin{aligned} \Gamma : \mathbb{F}_{q^m} &\to \mathbb{F}_q^m, \\ \sum_{i=1}^{m} \gamma_i x_i &\mapsto (x_1, \dots, x_m). \end{aligned}$$

By extending this map component-wise to vectors, i.e., $\Gamma : \mathbb{F}_{q^m}^n \to \mathbb{F}_q^{m \times n}$, we can equivalently define the rank weight of $\boldsymbol{x}$ as the rank of $\Gamma(\boldsymbol{x})$. This gives rise to two types of support: the row support is the row span of $\Gamma(\boldsymbol{x})$ which is a subspace in $\mathbb{F}_q^n$ and the column support is the column span of $\Gamma(\boldsymbol{x})$ which is in $\mathbb{F}_q^m$. In the following, $\text{supp}(\boldsymbol{x})$ will denote the row support unless specified otherwise.

Erasure Decoding. Given a code $\mathcal{C}$, the *error decoding* problem is to recover $\boldsymbol{c} \in \mathcal{C}$ given the received word $\boldsymbol{r} = \boldsymbol{c} + \boldsymbol{e}$ corrupted by an unknown error vector $\boldsymbol{e}$ of weight w. This becomes an *erasure decoding* problem if we also know the support of this error. In the Hamming metric, this means we know the location of the nonzero entries of $\boldsymbol{e}$ but not its values. In the rank metric, this means we know the row (or column) support of $\boldsymbol{e}$.

Error-Erasure Decoding. Suppose $\boldsymbol{r} = \boldsymbol{c} + \boldsymbol{e}' + \boldsymbol{e}$ where we have two unknown errors $\boldsymbol{e}'$ and $\boldsymbol{e}$ of weights w' and w, respectively. If we are also given the support of $\boldsymbol{e}'$ then this becomes an error-erasure decoding problem with erasure weight w' and error weight w. Many codes have efficient error-erasure decoders; for instance, Reed-Solomon codes and Gabidulin codes can correct w' erasures and w errors whenever $w' + 2w < d$, where d is the minimum distance of the code.

3 The PKE Framework

We keep the metric unspecified, as one can use either the Hamming or rank metric.

Algorithm. Fix a generator matrix $\mathbf{G} \in \mathbb{F}_{q^m}^{k \times n}$ of a code $\mathcal{C}$ that has an efficient error-erasure decoder $\textsf{error-erasure-dec}(\cdot)$. Let w' and w be large and small integers, respectively. This code, its decoder, and w' and w are public knowledge.

Key Generation. (Algorithm 1) Alice draws random vectors $\boldsymbol{m}' \in \mathbb{F}_{q^m}^k$ and $\boldsymbol{e}' \in \mathbb{F}_{q^m}^n$ such that $\text{wt}(\boldsymbol{e}') = w'$. The public key is $\boldsymbol{\rho} = \boldsymbol{m}'\mathbf{G} + \boldsymbol{e}'$, i.e., a random codeword of $\mathcal{C}$ corrupted by some large error. The secret key is $\boldsymbol{\kappa} = \boldsymbol{e}'$.

Encryption. (Algorithm 2) Given a message $\boldsymbol{m} \in \mathbb{F}_{q^m}^k$, Bob picks a random $\alpha \in \mathbb{F}_{q^m}^*$ and a random vector $\boldsymbol{e} \in \mathbb{F}_{q^m}^n$ of weight w. The ciphertext is given by $\boldsymbol{r} = \boldsymbol{m}\mathbf{G} + \alpha\boldsymbol{\rho} + \boldsymbol{e}$.

Decryption. (Algorithm 3) Note that the ciphertext $\boldsymbol{r}$ is a corrupted codeword of $\mathcal{C}$ with error $\alpha\boldsymbol{e}' + \boldsymbol{e}$. Since multiplication with the scalar α preserves the support of $\boldsymbol{e}'$[1], Alice knows $\text{supp}(\boldsymbol{e}') = \text{supp}(\alpha\boldsymbol{e}')$ and can perform error-erasure

[1] Note this holds for the Hamming metric and, assuming we use the concept of row support, also for the rank metric.

decoding to obtain $\alpha \boldsymbol{e}' + \boldsymbol{e}$. With $\alpha \boldsymbol{e}' + \boldsymbol{e}$ in hand, it is possible to extract α and hence find $\boldsymbol{m}$. Concrete examples of this extraction step in the Hamming and rank metric are given after the next paragraph.

Algorithm 1: Keygen

Input: —
Output: public key $\boldsymbol{\rho} \in \mathbb{F}_{q^m}^n$, secret key $\boldsymbol{\kappa} \in \mathbb{F}_{q^m}^n$

$\boldsymbol{e}' \xleftarrow{\$} \mathbb{F}_{q^m}^n$ s.t. $\mathrm{wt}(\boldsymbol{e}') = w'$
$\boldsymbol{m}' \xleftarrow{\$} \mathbb{F}_{q^m}^k$
$\boldsymbol{\rho} \leftarrow \boldsymbol{m}'\mathbf{G} + \boldsymbol{e}'$
$\boldsymbol{\kappa} \leftarrow \boldsymbol{e}'$
return $(\boldsymbol{\rho}, \boldsymbol{\kappa})$

Algorithm 2: Encryption

Input: message $\boldsymbol{m} \in \mathbb{F}_{q^m}^k$, Public key $\boldsymbol{\rho} \in \mathbb{F}_{q^m}^n$
Output: ciphertext $\boldsymbol{r} \in \mathbb{F}_{q^m}^n$

$\alpha \xleftarrow{\$} \mathbb{F}_{q^m}^*$
$\boldsymbol{e} \xleftarrow{\$} \mathbb{F}_{q^m}^n$ s.t. $\mathrm{wt}(\boldsymbol{e}) = w$
$\boldsymbol{r} \leftarrow \boldsymbol{m}\mathbf{G} + \alpha\boldsymbol{\rho} + \boldsymbol{e}$
return $\boldsymbol{r}$

Algorithm 3: Decryption

Input: ciphertext $\boldsymbol{r} \in \mathbb{F}_{q^m}^n$, secret key $\boldsymbol{\kappa} = \boldsymbol{e}' \in \mathbb{F}_{q^m}^n$
Output: message $\hat{\boldsymbol{m}} \in \mathbb{F}_{q^m}^k$

$\boldsymbol{x} \leftarrow$ error-erasure-dec $(\boldsymbol{r}, \mathrm{supp}(\boldsymbol{e}'))$ // $\boldsymbol{x}$ should be $\alpha\boldsymbol{e}' + \boldsymbol{e}$
$\hat{\alpha} \leftarrow$ extract α from $\boldsymbol{x}$ // see examples below
Solve $\hat{\boldsymbol{m}}\mathbf{G} = \boldsymbol{r} - \boldsymbol{x} - \hat{\alpha}\boldsymbol{m}'\mathbf{G}$ for $\hat{\boldsymbol{m}}$ // $\boldsymbol{r} - \boldsymbol{x}$ should equal $(\boldsymbol{m} + \alpha\boldsymbol{m}')\mathbf{G}$
return $\hat{\boldsymbol{m}}$

Intuition. The code $\mathcal{C} = \langle \mathbf{G} \rangle$ has a publicly known decoder that can efficiently correct a certain number of errors and erasures. The public key $\boldsymbol{\rho}$ is a corrupted codeword of $\mathcal{C}$ with an error weight w' far above the decoder's error correction capability. The idea for encryption is as follows: Bob wants to add an error to $\boldsymbol{m}\mathbf{G}$ such that Alice can decode it but not anyone else. Since the public key $\boldsymbol{\rho}$ contains a large error of weight w' that is only known to Alice, one possibility is for Bob to add a random multiple of $\boldsymbol{\rho}$ to obtain the ciphertext $\tilde{\boldsymbol{r}} = \boldsymbol{m}\mathbf{G} + \alpha\boldsymbol{\rho}$. The error added is $\alpha\boldsymbol{e}'$ and its support is known only to Alice. Hence, for Alice, $\tilde{\boldsymbol{r}}$ contains only erasures while for an attacker $\tilde{\boldsymbol{r}}$ contains errors. However, this would not be secure as the ciphertext is just a codeword of the known extended code $\left\langle \begin{bmatrix} \mathbf{G} \\ \boldsymbol{\rho} \end{bmatrix} \right\rangle$ and thus $\begin{bmatrix} \boldsymbol{m} \\ \alpha \end{bmatrix}$ can be obtained by matrix inversion. To prevent this attack, Bob also adds a small weight-w error $\boldsymbol{e}$ of his own. Hence, the final ciphertext expression is $\boldsymbol{r} = \boldsymbol{m}\mathbf{G} + \alpha\boldsymbol{\rho} + \boldsymbol{e}$. Thus, if Alice can decode w' erasures and w errors in this code, then she can correctly decrypt. However, an attacker seemingly faces the problem of decoding at least w' errors. With a suitable code and parameter choice, one expects the former problem to be easy and the latter to be hard.

Example: Hamming Metric. Let us give more details for the decryption step.

The first step consists of error-erasure decoding the ciphertext $\boldsymbol{r}$ to obtain $\alpha\boldsymbol{e}' + \boldsymbol{e}$, while the second step extracts α from $\alpha\boldsymbol{e}' + \boldsymbol{e}$. After this, Alice can easily find $\boldsymbol{m}$.

- We illustrate error-erasure decoding by assuming the code is an RS code. Since
$$\boldsymbol{r} = \boldsymbol{m}\mathbf{G} + \alpha\boldsymbol{\rho} + \boldsymbol{e} = (\boldsymbol{m} + \alpha\boldsymbol{m}')\mathbf{G} + (\alpha\boldsymbol{e}' + \boldsymbol{e}),$$
if Alice punctures $\boldsymbol{r}$ at $\mathrm{supp}(\boldsymbol{e}')$, then the $\boldsymbol{\alpha e}'$ term disappears and she is left with a corrupted codeword of error weight at most w in a punctured RS code of blocklength $n - w'$, which is still an RS code. Thus, if $w \leq \frac{n-w'-k}{2}$, she can decode to obtain $\boldsymbol{m} + \alpha\boldsymbol{m}'$ or equivalently $\alpha\boldsymbol{e}' + \boldsymbol{e}$.
- Since $w < w'$, several entries of $\alpha\boldsymbol{e}' + \boldsymbol{e}$ must be equal to $\alpha e_i'$ and hence reveal α.
- Since Alice knows $\boldsymbol{m}'$, α and $(\boldsymbol{m} + \alpha\boldsymbol{m}')$, she can find $\boldsymbol{m}$ (step 3 of Algorithm 3).

Example: Rank Metric. We again explain the decryption in more detail, where the first two steps consist of error-erasure decoding and extraction.

- We illustrate error-erasure decoding assuming a Gabidulin code is used. Since Alice knows the row support of $\boldsymbol{e}'$, she knows a matrix $\mathbf{J} \in \mathbb{F}_q^{n\times n}$ such that the vector $\boldsymbol{e}'\mathbf{J} \in \mathbb{F}_{q^m}^n$ is zero in the first $n - w'$ positions. Note that
$$\boldsymbol{r}\mathbf{J} = \boldsymbol{m}\mathbf{G}\mathbf{J} + \alpha\boldsymbol{\rho}\mathbf{J} + \boldsymbol{e}\mathbf{J} = (\boldsymbol{m} + \alpha\boldsymbol{m}')\mathbf{G}\mathbf{J} + (\alpha\boldsymbol{e}'\mathbf{J} + \boldsymbol{e}\mathbf{J}).$$
Thus, if we puncture in the last w' positions, the $\alpha\boldsymbol{e}'\mathbf{J}$ term disappears and we are left with an error-only decoding problem with error weight equal to $\mathrm{rank}_{\mathbb{F}_q}(\boldsymbol{e}\mathbf{J}) \leq w$. This decoding problem is in a code of blocklength $n - w'$, obtained by puncturing the last w' positions of $\langle\mathbf{GJ}\rangle$. This is still a Gabidulin code and, if $w \leq \frac{n-w'-k}{2}$, Alice can decode to obtain $\alpha\boldsymbol{e}' + \boldsymbol{e}$.
- Extracting α can be done as follows. The error-erasure decoder outputs $\boldsymbol{x} = \alpha\boldsymbol{e}' + \boldsymbol{e}$. Multiplying this with the same $\mathbf{J}$ as above, Alice gets $\boldsymbol{x}\mathbf{J} = \boldsymbol{e}'\mathbf{J} + \boldsymbol{e}\mathbf{J}$. Note the first $n - w'$ positions of $\boldsymbol{x}\mathbf{J}$ equal $\boldsymbol{e}\mathbf{J}$ and with high probability the $\mathbb{F}_q$-span of these $n - w'$ positions will equal $\langle \boldsymbol{e}\rangle_{\mathbb{F}_q}$. Alice picks any $\boldsymbol{a} \in \mathbb{F}_{q^m}^w$ such that $\langle \boldsymbol{a}\rangle_{\mathbb{F}_q} = \langle \boldsymbol{e}\rangle_{\mathbb{F}_q}$. She can now write $\boldsymbol{x} = \alpha\boldsymbol{e}' + \boldsymbol{a}\boldsymbol{B}$ where only $\alpha \in \mathbb{F}_{q^m}^*$ and $\boldsymbol{B} \in \mathbb{F}_q^{w\times n}$ are unknown. This is a linear system of mn equations over $\mathbb{F}_q$ with $m + nw$ unknowns. Since w is small, this system is overdetermined and hence has a unique solution with high probability.
- As before, since Alice knows $\boldsymbol{m}'$, α and $(\boldsymbol{m} + \alpha\boldsymbol{m}')$, she can find $\boldsymbol{m}$.

The framework can also be used in the rank metric using the concept of column support instead. In this case, Bob needs to multiply $\boldsymbol{\rho}$ by an α drawn from $\mathbb{F}_q^*$ instead of $\mathbb{F}_{q^m}^*$, in order to keep the column support unchanged. Then Alice can decrypt assuming she has an error-erasure decoder, with erasures here referring to knowing the column support.

4 Security and Correctness

The following assertions are easy to show.

Correctness. The framework is correct if the code family has an error-erasure decoder that can efficiently correct w errors and w' erasures. In the Hamming metric, decryption is always successful, while in the rank metric, we tolerate a small decryption failure probability since the extraction step can fail sometimes.

Security. The framework has one-wayness under chosen plaintext attack (OW-CPA) security if the following problem is hard. Let $\mathcal{S}_w$ denote the set of vectors in $\mathbb{F}_{q^m}^n$ of weight w. The problem is parameterized by a public generator matrix $\mathbf{G} \in \mathbb{F}_{q^m}^{k \times n}$ and integers w', w satisfying $w' > w$.

Problem 1 (Supercode Decoding $(\mathbf{G}, w, w')$*).* Draw $\boldsymbol{c'} \in \langle \mathbf{G} \rangle$ and $\boldsymbol{e'} \in \mathcal{S}_{w'}$ uniformly at random. Set $\boldsymbol{\rho} = \boldsymbol{c'} + \boldsymbol{e'}$. Draw $\boldsymbol{c}^{\dagger} \in \left\langle \begin{bmatrix} \mathbf{G} \\ \boldsymbol{\rho} \end{bmatrix} \right\rangle \setminus \langle \mathbf{G} \rangle$ and $\boldsymbol{e} \in \mathcal{S}_w$ uniformly at random. Set $\boldsymbol{r} = \boldsymbol{c}^{\dagger} + \boldsymbol{e}$. Given only $\mathbf{G}$, $\boldsymbol{\rho}$ and $\boldsymbol{r}$, find $\boldsymbol{e}$.

An open question, then, is whether there exists a code and integers w', w such that the above assertions hold simultaneously.

5 A Survey of Past Schemes

Our framework is inspired by and generalizes the Augot-Finiasz and Faure-Loidreau systems. However, this generalization is not exact and these proposals exhibit some differences. To better understand these past schemes, we introduce a *shorter message* and an *interleaved* variation of the framework.

5.1 Shorter Message Variation

In this variation, the first entry of the message $\boldsymbol{m}$ is required to be 0 while the first entry of $\boldsymbol{m'}$ is nonzero. This simplifies the decryption algorithm: after the error-erasure decoder returns $\alpha \boldsymbol{e'} + \boldsymbol{e}$, or equivalently $\alpha \boldsymbol{m'} + \boldsymbol{m}$, we can recover α directly by comparing the first entry of $\alpha \boldsymbol{m'} + \boldsymbol{m}$ with the first entry of $\boldsymbol{m'}$. In specific cases, the shorter message variation has benefits. In the case of rank metric, there is no longer any decryption failure probability, and for the interleaved variant in the Hamming metric, this shorter message technique is even required to make decryption work. The big drawback, in our view, is that the underlying hard problem becomes more ad hoc and complex.

Algorithm 4: Keygen (Interleaving)

Input: —
Output: public key $\boldsymbol{\rho} \in (\mathbb{F}_{q^m}^n)^\ell$, secret key $\boldsymbol{\kappa} \in (\mathbb{F}_{q^m}^n)^\ell$

$\boldsymbol{e}'^{(1)}, \ldots, \boldsymbol{e}'^{(\ell)} \xleftarrow{\$} \mathbb{F}_{q^m}^n$
s.t. $|\bigcup_{i=1}^{\ell} \operatorname{supp}(\boldsymbol{e}'^{(i)})| = w'$ if Hamming metric
or $|\operatorname{supp}(\boldsymbol{e}'^{(1)}) + \cdots + \operatorname{supp}(\boldsymbol{e}'^{(\ell)})| = w'$ if rank metric
$\boldsymbol{m}'^{(1)}, \ldots, \boldsymbol{m}'^{(\ell)} \xleftarrow{\$} \mathbb{F}_{q^m}^k$
$\boldsymbol{\rho} \leftarrow (\boldsymbol{\rho}^{(1)}, \ldots, \boldsymbol{\rho}^{(\ell)})$ where $\boldsymbol{\rho}^{(i)} \leftarrow \boldsymbol{m}'^{(i)}\mathbf{G} + \boldsymbol{e}'^{(i)}$
$\boldsymbol{\kappa} \leftarrow (\boldsymbol{e}'^{(1)}, \ldots, \boldsymbol{e}'^{(\ell)})$
return $(\boldsymbol{\rho}, \boldsymbol{\kappa})$

Algorithm 5: Encryption (Interleaving)

Input: message $\boldsymbol{m} \in \mathbb{F}_{q^m}^k$, public key $\boldsymbol{\rho} \in (\mathbb{F}_{q^m}^n)^\ell$
Output: ciphertext $\boldsymbol{r} \in \mathbb{F}_{q^m}^n$

$\alpha_1, \ldots, \alpha_\ell \xleftarrow{\$} \mathbb{F}_{q^m}$ not all zero
$\boldsymbol{e} \xleftarrow{\$} \mathbb{F}_{q^m}^n$ s.t. $\operatorname{wt}(\boldsymbol{e}) = w$
$\boldsymbol{r} \leftarrow \boldsymbol{m}\mathbf{G} + \sum_{i=1}^{\ell} \alpha_i \boldsymbol{\rho}^{(i)} + \boldsymbol{e}$
return $\boldsymbol{r}$

5.2 Interleaved Variation

Key Generation. (Algorithm 4) In this variant, the public key consists of ℓ corrupted codewords, where ℓ is called the interleaving order. That is, $\boldsymbol{\rho} = (\boldsymbol{\rho}^{(1)}, \ldots, \boldsymbol{\rho}^{(\ell)})$ where $\boldsymbol{\rho}^{(i)} = \boldsymbol{m}'^{(i)}\mathbf{G} + \boldsymbol{e}'^{(i)}$ for $i = 1, \ldots, \ell$. We choose $\boldsymbol{m}'^{(i)}$ independently and uniformly at random. The error vectors $\boldsymbol{e}'^{(i)}$ are random weight-w' vectors but with the additional constraint that their support is confined to a common support.[2] In the Hamming metric, this means $|\bigcup_{i=1}^{\ell} \operatorname{supp}(\boldsymbol{e}'^{(i)})| = w'$ while the condition in the rank metric becomes $|\operatorname{supp}(\boldsymbol{e}'^{(i)}) + \cdots + \operatorname{supp}(\boldsymbol{e}'^{(\ell)})| = w'$.[3]

Encryption. (Algorithm 5) Bob chooses random $\alpha_i \in \mathbb{F}_{q^m}$, not all zero, for $i = 1, \ldots, \ell$ and an error vector $\boldsymbol{e}$ of weight w. The ciphertext is then given by $\boldsymbol{r} = \boldsymbol{m}\mathbf{G} + \sum_{i=1}^{\ell} \alpha_i \boldsymbol{\rho}^{(i)} + \boldsymbol{e}$. That is, Bob adds a random non-zero linear combination of the $\boldsymbol{\rho}^{(i)}$'s and a small weight error vector.

Decryption. (Algorithm 6) Let us first consider the rank metric. Alice knows a w'-dimensional space containing the support of all the errors $\boldsymbol{e}'^{(i)}$ and thus,

[2] Actually, the key generation could be modified so that the $\boldsymbol{e}'^{(i)}$'s are also linearly independent over $\mathbb{F}_{q^m}$, since this is necessary for the extraction step during decryption. But, as we will see later, this is only relevant in the rank metric since extraction is inefficient in the Hamming metric anyway.

[3] Recall that, for the rank metric, $\operatorname{supp}(\boldsymbol{e}'^{(i)}) \subseteq \mathbb{F}_q^n$ is the row support of $\boldsymbol{e}'$ and $|\mathcal{U}|$ denotes the dimension of the subspace $\mathcal{U}$.

Algorithm 6: Decryption (Interleaving)

Input: ciphertext $\boldsymbol{r} \in \mathbb{F}_{q^m}^n$, secret key $\boldsymbol{\kappa} = (\boldsymbol{e}'^{(1)}, \ldots, \boldsymbol{e}'^{(\ell)}) \in (\mathbb{F}_{q^m}^n)^\ell$
Output: message $\hat{\boldsymbol{m}} \in \mathbb{F}_{q^m}^k$

$\boldsymbol{x} \leftarrow$ error-erasure-dec $(\boldsymbol{r}, \cup_{i=1}^{\ell} \operatorname{supp}(\boldsymbol{e}'^{(i)}))$ if Hamming metric
or $\boldsymbol{x} \leftarrow$ error-erasure-dec $(\boldsymbol{r}, \operatorname{supp}(\boldsymbol{e}'^{(1)}) + \cdots + \operatorname{supp}(\boldsymbol{e}'^{(\ell)}))$ if rank metric
// $\boldsymbol{x}$ should be $\sum_{i=1}^{\ell} \alpha_i \boldsymbol{e}'^{(i)} + \boldsymbol{e}$

$\hat{\alpha}_1, \ldots, \hat{\alpha}_\ell \leftarrow$ extract $\alpha_1, \ldots, \alpha_\ell$ from $\boldsymbol{x}$
// inefficient in Hamming metric

$\hat{\boldsymbol{m}} \leftarrow$ solve $\hat{\boldsymbol{m}}\mathbf{G} = \boldsymbol{r} - \boldsymbol{x} - \sum_{i=1}^{\ell} \hat{\alpha}_i \boldsymbol{m}'^{(i)}\mathbf{G}$ for $\hat{\boldsymbol{m}}$
// $\boldsymbol{r} - \boldsymbol{x}$ should equal $(\boldsymbol{m} + \sum_{i=1}^{\ell} \alpha_i \boldsymbol{m}'^{(i)})\mathbf{G}$

return $\hat{\boldsymbol{m}}$

in particular, the support of $\sum_{i=1}^{\ell} \alpha_i \boldsymbol{e}'^{(i)}$. Hence, she can obtain $\boldsymbol{e} + \sum_i \alpha_i \boldsymbol{e}'^{(i)}$ by error-erasure decoding $\boldsymbol{r}$. Then, using an extraction step similar to the non-interleaved case, she can find $\alpha_1, \ldots, \alpha_\ell$ with high probability and hence $\boldsymbol{m}$.

The Hamming metric case is similar: Alice knows a w'-size set of indices containing the support of all the errors $\boldsymbol{e}'^{(i)}$ and thus, in particular, the support of $\sum_{i=1}^{\ell} \alpha_i \boldsymbol{e}'^{(i)}$. Hence, she can obtain $\boldsymbol{e} + \sum_i \alpha_i \boldsymbol{e}'^{(i)}$ by error-erasure decoding $\boldsymbol{r}$. However, there does not seem to be an efficient way to extract $\alpha_1, \ldots, \alpha_\ell$ from $\boldsymbol{e} + \sum_i \alpha_i \boldsymbol{e}'^{(i)}$ unless we use the shorter message variation described next.

Shorter Message Variation with Interleaving. With this variation, the first ℓ entries of $\boldsymbol{m}$ are required to be 0 and the first ℓ entries of each $\boldsymbol{m}'^{(i)}$ are set to zero except the ith entry which is set to 1.[4] Thus, the first ℓ entries of $\boldsymbol{m} + \sum_{i=1}^{\ell} \alpha_i \boldsymbol{m}'^{(i)}$ equal $(\alpha_1, \ldots, \alpha_\ell)$ and hence finding $\boldsymbol{m}$ is straightforward after the error-erasure decoder returns $\boldsymbol{m} + \sum_{i=1}^{\ell} \alpha_i \boldsymbol{m}'^{(i)}$.

Intuitively, the interleaved version might be harder to break since there are more unknowns α_i in the ciphertext. However, as in the shorter message variation, the big drawback is that the system and the underlying hard problem (Appendix A) becomes more complex to analyse.

5.3 Framework with RS and Gabidulin Codes

We describe our framework instantiated with RS and Gabidulin codes and explain why these choices are insecure. Many of the previously proposed schemes and the attacks against them can be seen in this framework. Since several of these schemes also use the shorter message and/or the interleaved variation, we also account for them.

---

[4] This can be generalized to the requirement that the vectors $\boldsymbol{\rho}_{1:\ell}^{(1)}, \boldsymbol{\rho}_{1:\ell}^{(2)}, \ldots, \boldsymbol{\rho}_{1:\ell}^{(\ell)}$ are linearly independent where $\boldsymbol{x}_{1:\ell}$ denotes the first ℓ entries of the vector $\boldsymbol{x}$.

Remark 1. The length of the message vector $\boldsymbol{m}$ is k, but for the past schemes that use the shorter message variation, the first ℓ entries will be set to zero (where ℓ denotes the interleaving order and $\ell = 1$ in the non-interleaved case). We will denote this "effective length" of $\boldsymbol{m}$ with k'. We have $k' = k - \ell$ if we use the shorter message variation, else $k' = k$.

Assume that we instantiate the scheme in the Hamming metric using an RS code with evaluation points $\boldsymbol{a} = (a_1, \ldots, a_n)$. Recall that the public key and the ciphertext are given by

$$\boldsymbol{\rho} = \boldsymbol{m}'\mathbf{G} + \boldsymbol{e}', \tag{1}$$

$$\boldsymbol{r} = \boldsymbol{m}\mathbf{G} + \alpha\boldsymbol{\rho} + \boldsymbol{e}. \tag{2}$$

The linearization attack due to Coron [6] works as follows:[5] let $V_0 \in \mathcal{P}_{\leq w}$ be a polynomial of degree w such that for all i we have $V_0(a_i) \cdot e_i = 0$, i.e., $V_0(\boldsymbol{a}) \cdot \boldsymbol{e} = \mathbf{0}$. Then, if we component-wise multiply (2) by $V_0(\boldsymbol{a})$, we get

$$V_0(\boldsymbol{a}) \cdot \boldsymbol{r} = V_0(\boldsymbol{a}) \cdot m(\boldsymbol{a}) + V_0(\boldsymbol{a}) \cdot \alpha\boldsymbol{\rho} \tag{3}$$

where $m \in \mathcal{P}_{<k'}$ is the polynomial such that $m(\boldsymbol{a}) = \boldsymbol{m}\mathbf{G}$. We now linearize this system of equations. Let $V \in \mathcal{P}_{\leq w}, N \in \mathcal{P}_{<w+k'}$ and $R \in \mathcal{P}_{\leq w}$ be polynomials such that

$$V(\boldsymbol{a}) \cdot \boldsymbol{r} = N(\boldsymbol{a}) + R(\boldsymbol{a}) \cdot \boldsymbol{\rho}. \tag{4}$$

The first set of Eq. (3) has a non-zero solution (V_0, m, α) by the problem statement. Since the second set of equations (4) is more general than the first (3), this non-zero solution also yields a solution $(V = V_0, N = V_0 \cdot m, R = \alpha V_0)$ of (4). The second system of equations is homogeneous and linear, and since it has a non-zero solution, its kernel has dimension at least one. If we assume this dimension to be exactly one, then the solution space is generated by (V_0, m, α) and so (3) can be solved by solving (4). The latter has $(w+1)+(w+k')+(w+1) = k'+3w+2$ unknowns and n equations. If $n \geq k' + 3w + 1$, then the kernel has with high probability dimension one, as observed experimentally in [6]. Thus, the system can be broken.

Now we consider the generalization of this attack to the interleaved variant. Here the public key and ciphertext expressions are

$$\boldsymbol{\rho}^{(i)} = \boldsymbol{m}'^{(i)}\mathbf{G} + \boldsymbol{e}'^{(i)} \quad \text{for } i = 1, \ldots, \ell, \tag{5}$$

$$\boldsymbol{r} = \boldsymbol{m}\mathbf{G} + \sum_{i=1}^{\ell} \alpha_i \boldsymbol{\rho}^{(i)} + \boldsymbol{e}. \tag{6}$$

[5] In this non-interleaved case, Coron actually derived an improved attack using determinants that did not involve linearizing the αV term. This had the advantage that his attack was not probabilistic and had a more relaxed requirement on the parameters for the attack to work.

Then, as before, we can set up the following two systems of equations. Let $V_0 \in \mathcal{P}_{\le w}$ be such that

$$V_0(\boldsymbol{a}) \cdot \boldsymbol{r} = V_0(\boldsymbol{a}) \cdot m(\boldsymbol{a}) + V_0(\boldsymbol{a}) \cdot \sum_{j=1}^{\ell} \alpha_j \boldsymbol{\rho}^{(j)}. \tag{7}$$

Let $V \in \mathcal{P}_{\le w}, N \in \mathcal{P}_{<w+k'}$ and $R_1, \dots, R_\ell \in \mathcal{P}_{\le w}$ be such that

$$V(\boldsymbol{a}) \cdot \boldsymbol{r} = N(\boldsymbol{a}) + \sum_{j=1}^{\ell} R_\ell(\boldsymbol{a}) \cdot \boldsymbol{\rho}^{(j)}. \tag{8}$$

The latter system has $(w+1)+(w+k')+\ell(w+1) = (\ell+2)w+k'+\ell+1$ unknowns and n equations. As before, this system will have a kernel of dimension one with high probability if $n \ge (\ell+2)w + k' + \ell$.

The attack in the rank metric using Gabidulin codes is similar, where linearized polynomials are used instead. Let $\boldsymbol{g} = (g_1, \dots, g_n) \in \mathbb{F}_{q^m}^n$ be the evaluation points of the Gabidulin code, where $g_1, \dots, g_n$ are linearly independent over $\mathbb{F}_q$. Let $V_0 \in \mathcal{L}_{\le w}$ be a q-polynomial of q-degree w such that for all i we have $V_0(e_i) = 0$, i.e., $V_0(\boldsymbol{e}) = \boldsymbol{0}$. Such a V_0 exists since, for any given $\mathbb{F}_q$-subspace of $\mathbb{F}_{q^m}$ of dimension w, it is always possible to find a q-polynomial of q-degree w whose kernel equals this subspace [14, Theorem 7]. We have

$$V_0(\boldsymbol{r}) = V_0 \circ m(\boldsymbol{g}) + \sum_{j=1}^{\ell} V_0(\alpha_j \boldsymbol{\rho}^{(j)}) \tag{9}$$

where $m \in \mathcal{L}_{<k'}$ is the q-polynomial such that $m(\boldsymbol{g})$ equals $\boldsymbol{m}\mathbf{G}$. Let $V \in \mathcal{L}_{\le w}$, $N \in \mathcal{L}_{<w+k'}$ and $R \in \mathcal{L}_{\le w}$ be such that

$$V(\boldsymbol{r}) = N(\boldsymbol{g}) + \sum_{j=1}^{\ell} R_\ell(\boldsymbol{\rho}^{(j)}). \tag{10}$$

For a solution of the second system to yield a solution to the first, we again require $n \ge (\ell+2)w + k' + \ell$.

Is it possible to choose parameters such that $n < (\ell+2)w + k' + \ell$ in order to thwart this attack? The obvious solution of arbitrarily increasing ℓ does not work. It is well-known that interleaved Reed-Solomon codes (see, e.g., [4,18]) and interleaved Gabidulin codes (see [13,19] for common column support and [15, Section 3.2] for common row support) can be decoded with high probability up to $\frac{\ell}{\ell+1}(n-k)$ errors, where the number of "errors" here refers to the size/dimension of the common support shared by these ℓ error vectors. Thus, if $w' \le \frac{\ell}{\ell+1}(n-k)$ then it becomes possible to recover the private key $(\boldsymbol{e}'^{(1)}, \dots, \boldsymbol{e}'^{(\ell)})$ from the public key $(\boldsymbol{\rho}^{(1)}, \dots, \boldsymbol{\rho}^{(\ell)})$. As a result, we have the following conditions on the cryptosystem:

$$
\begin{aligned}
2w + w' &\leq n - k && \text{for correct decryption,} \quad (11)\\
n &< (\ell + 2)w + k' + \ell && \text{to prevent linearization attack,} \quad (12)\\
w' &> \frac{\ell}{\ell + 1}(n - k) && \text{to prevent interleaved decoding attack.} \quad (13)
\end{aligned}
$$

It can be shown that these inequalities are incompatible, see Appendix B, and thus the system is insecure with both RS and Gabidulin codes.

5.4 Past Schemes

Augot-Finiasz Scheme (2003). [2,3]. This scheme is obtained by instantiating our framework with RS codes in the Hamming metric. The only difference is that Augot and Finiasz use the shorter message variant (which, in this case, provides no advantage in our view). In a subsequent work [3], they introduce the interleaved version of their scheme, which they do so in terms of the trace operator. Both were successfully attacked by Coron's linearization attack [6] in 2004, which we explained in Sect. 5.3. Coron also showed that increasing the interleaving order ℓ cannot help by describing an algorithm, which is equivalent to an interleaved RS decoder, that recovers the private key. Kiayias and Yung [11] showed that the non-interleaved version of the system is still insecure if we increase w by (a) using Guruswami-Sudan list decoding in the decryption step instead of unique decoding and (b) using the fact that some of the w errors will overlap with the w' errors in the public key and thus a larger w can still allow decryption (with a small probability of failure).

Faure-Loidreau Scheme (2006). [7] This is the rank-metric counterpart of the Augout-Finiasz scheme using Gabidulin codes. The paper noted that the non-interleaved version is vulnerable to a Coron-like linearization attack and put forward the interleaved variant with a large ℓ to avoid this attack. As the discussion in Sect. 5.3 shows, if the interleaving order ℓ is chosen large enough to resist the linearization attack, it becomes vulnerable to an interleaved decoding attack. Thus, this scheme was also broken, and this vulnerability was shown by Gaborit et al. [9].

RAMESSES (Lavauzelle et al., 2019). [12] This scheme was proposed as a repair of the non-interleaved variant of the Faure-Loidreau cryptosystem. A new feature of their scheme is that they preserve the column support of the error vector $\boldsymbol{e}'$ instead of the row support. They formulate the scheme using the parity-check matrix but we will give an equivalent formulation using the generator matrix to keep our presentation unified. We note the scheme sets $n = m$.

Consider the ciphertext expression in the original Faure-Loidreau system,

$$
\boldsymbol{r} = \boldsymbol{m}\mathbf{G} + \alpha\boldsymbol{\rho} + \boldsymbol{e}
$$

where $\alpha \in \mathbb{F}_{q^m}^*$ and $\boldsymbol{e} \in \mathbb{F}_{q^m}^n$ of weight w are drawn uniformly at random. To preserve the column support of $\boldsymbol{e}'$ in $\boldsymbol{\rho}$, we need to draw α from $\mathbb{F}_q^*$ instead of $\mathbb{F}_{q^m}^*$. RAMESSES then goes further and also allows random column operations over $\mathbb{F}_q$ on the vector $\boldsymbol{\rho}$, i.e., right-multiplying $\boldsymbol{\rho}$ by a random invertible matrix $\mathbf{T} \in \mathbb{F}_q^{m \times m}$. Hence the ciphertext becomes

$$\boldsymbol{r} = \boldsymbol{m}\mathbf{G} + \boldsymbol{\rho}\mathbf{T} + \boldsymbol{e}. \tag{14}$$

Notice we do not need α anymore since its effect can be subsumed in $\mathbf{T}$. However, this poses a difficulty for Alice to decrypt. The ciphertext can be written as

$$\boldsymbol{r} = \boldsymbol{m}\mathbf{G} + \boldsymbol{\rho}\mathbf{T} + \boldsymbol{e} = \boldsymbol{m}\mathbf{G} + \boldsymbol{m}'\mathbf{G}\mathbf{T} + \boldsymbol{e}'\mathbf{T} + \boldsymbol{e}. \tag{15}$$

The $\boldsymbol{m}\mathbf{G}$ term is the codeword, the $\boldsymbol{e}'\mathbf{T}$ term is an erasure for Alice since she knows its column support (of dimension w') and the $\boldsymbol{e}$ term is an error (of rank w). But $\boldsymbol{m}'\mathbf{G}\mathbf{T}$ is not necessarily a codeword of $\langle \mathbf{G} \rangle$ and it introduces a potentially large error, thus making $\boldsymbol{r}$ undecodeable for Alice.

To address this, we only use matrices $\mathbf{T}$ such that $\boldsymbol{g}\mathbf{T} \in \langle \boldsymbol{g}, \boldsymbol{g}^q, \ldots, \boldsymbol{g}^{q^\gamma} \rangle_{\mathbb{F}_{q^m}}$, where γ is another public parameter of the cryptosystem and $\boldsymbol{g} \in \mathbb{F}_{q^m}^n$ is the list of evaluation points for the Gabidulin code. Then, for all codewords $\boldsymbol{c}$ of this $[n, k]$ Gabidulin code, $\boldsymbol{c}\mathbf{T}$ is in the bigger $[n, k+\gamma]$ Gabidulin code defined using the same evaluation points $\boldsymbol{g}$. In other words, $\boldsymbol{c}\mathbf{T} \in \left\langle \boldsymbol{g}, \boldsymbol{g}^q, \ldots, \boldsymbol{g}^{q^{k+\gamma}} \right\rangle_{\mathbb{F}_{q^m}}$. This follows because, if $\boldsymbol{c} = m(\boldsymbol{g})$ where $m \in \mathcal{L}_{<k}$, then $\boldsymbol{c}\mathbf{T} = m(\boldsymbol{g})\mathbf{T} = m(\boldsymbol{g}\mathbf{T})$ which is in $\left\langle \boldsymbol{g}, \boldsymbol{g}^q, \ldots, \boldsymbol{g}^{q^{k+\gamma}} \right\rangle_{\mathbb{F}_{q^m}}$. Alice can then decrypt by error-erasure decoding in this bigger code if

$$2w + w' \leq n - (k + \gamma). \tag{16}$$

The scheme was broken by Bombar and Couvreur [5] where they show how to adapt Coron's linearization attack. Consider the ciphertext expression again:

$$\boldsymbol{r} = \boldsymbol{m}\mathbf{G} + \boldsymbol{\rho}\mathbf{T} + \boldsymbol{e}.$$

The attack begins by rewriting this using q-polynomials. Each vector $\boldsymbol{x} \in \mathbb{F}_{q^m}^m$ corresponds to a unique q-polynomial $f \in \mathcal{L}_{<m}$ such that $\boldsymbol{x} = (f(g_1), \ldots, f(g_m))$, and this can be found using interpolation. Note addition in $\mathbb{F}_{q^m}^m$ corresponds to addition in $\mathcal{L}_{<m}$ under this mapping. Thus, we can write (14) as

$$r = m + j + e$$

where $m \in \mathcal{L}_{<k}$ and $r, e \in \mathcal{L}_{<m}$ are the q-polynomials found by interpolating $\boldsymbol{m}\mathbf{G}, \boldsymbol{r}$ and $\boldsymbol{e}$ respectively, and the q-polynomial j is the interpolated version of the vector $\boldsymbol{\rho}\mathbf{T}$. We will now find j. Let $\rho \in \mathcal{L}_{<m}$ be the interpolated version of $\boldsymbol{\rho}$. Let $T \in \mathcal{L}_{\leq \gamma}$ be the q-polynomial such that $T(\boldsymbol{g}) = \boldsymbol{g}\mathbf{T}$. The latter is in $\langle \boldsymbol{g}, \boldsymbol{g}^q, \ldots, \boldsymbol{g}^{q^\gamma} \rangle_{\mathbb{F}_{q^m}}$ and hence T has q-degree γ.[6] We have $\rho \circ T(\boldsymbol{g}) = \rho(\boldsymbol{g}\mathbf{T}) =$

[6] If $\mathbf{T}$ is chosen such that $\boldsymbol{g}\mathbf{T} \in \left\langle \boldsymbol{g}, \boldsymbol{g}^q, \ldots, \boldsymbol{g}^{q^\gamma} \right\rangle_{\mathbb{F}_{q^m}} \setminus \left\langle \boldsymbol{g}, \boldsymbol{g}^q, \ldots, \boldsymbol{g}^{q^{\gamma-1}} \right\rangle_{\mathbb{F}_{q^m}}$, then its q-degree is γ otherwise it is at most γ.

$\rho(\boldsymbol{g})\mathbf{T} = \boldsymbol{\rho}\mathbf{T}$. Thus $\rho \circ T$ is the interpolated version of $\boldsymbol{\rho}\mathbf{T}$. Hence,

$$r = m + \rho \circ T + e \tag{17}$$

Note that if we left-compose (17) with a $V \in \mathcal{L}_{\leq w}$ such that $V \circ e = 0$ (such a V can be shown to exist) then we get $V \circ r = V \circ m + V \circ \rho \circ T$. Then to linearize $V \circ \rho \circ T$ we would need to replace this entire term with an unknown polynomial of a large degree since we cannot take advantage of the fact that ρ is known as it appears between the unknowns V and T. Thus, this introduces too many unknowns for the attack to work. Indeed, what is needed is a composition from the right, and this is what [5] shows how to do.

Let $V \in \mathcal{L}_{\leq w}$ be a q-polynomial of q-degree at most w such that $e \circ V = 0$. That such a non-zero V exists is proved in [5, Proposition 1]. Thus,

$$r \circ V = m \circ V + \rho \circ T \circ V. \tag{18}$$

We can then linearize $m \circ V$ and $T \circ V$ to get

$$r \circ V = N + \rho \circ R \tag{19}$$

where $N \in \mathcal{L}_{<w+k}$ and $R \in \mathcal{L}_{\leq w+\gamma}$.

The unknowns here are the coefficients of V, N, and R. Note that this is not a linear system since $\rho \circ R$ contains higher degree terms in the coefficients of R. Nonetheless, [5] shows that this equation can be transformed to an equivalent system, which is linear, by using an adjoint operation for q-polynomials. Hence, the attack can be carried out.

Remark 2. The description above is *almost* equivalent to the original RAMESSES scheme, with the following differences. First, our scheme puts the plaintext in $\boldsymbol{m}$, whereas the original scheme places it in the row support of $\boldsymbol{e}$. Second, in our scheme, Bob adds the random codeword $\boldsymbol{m}\mathbf{G}$ which belongs to the $[n, k]$ Gabidulin code. However, the codeword can actually be drawn from the larger $[n, k + \gamma]$ Gabidulin code. With these two changes, our description becomes an equivalent "generator-matrix" version of the original scheme.

LIGA (Renner et al., 2021). [17] This scheme was another repair of the (interleaved variant of the) Faure-Loidreau system. We will first explain the idea of the repair and then show why it turned out to be insecure (shown by [5]). Recall that the public key is given by

$$\boldsymbol{\rho}^{(i)} = \boldsymbol{m}'^{(i)}\mathbf{G} + \boldsymbol{e}'^{(i)} \quad \text{where } i = 1, \ldots, \ell,$$

where the $\boldsymbol{e}'^{(i)}$'s are such that they share a common row support of dimension w'. That is,

$$\left|\operatorname{supp}(\boldsymbol{e}'^{(1)}) + \cdots + \operatorname{supp}(\boldsymbol{e}'^{(\ell)})\right| = w'. \tag{20}$$

As discussed in Sect. 5.3, if we choose the interleaving order ℓ large enough in order to avoid the linearization attack, the scheme becomes susceptible to an interleaved decoding attack. However, the paper [17] notes that all known interleaved decoders fail if a certain condition [17, Lemma 1] about the ℓ error vectors is satisfied. This latter condition, in turn, is implied by the following [17, Theorem 4],

$$\mathrm{rank}_{\mathbb{F}_{q^m}} \begin{bmatrix} \boldsymbol{e}'^{(1)} \\ \vdots \\ \boldsymbol{e}'^{(\ell)} \end{bmatrix} < \frac{w}{n-k-w}. \tag{21}$$

It is possible to pick $\boldsymbol{e}'^{(i)}$'s such that both (20) and (21) hold, and the repair suggests drawing the $\boldsymbol{e}'^{(i)}$'s in the key generation step in precisely this way.

This repair did not succeed because picking linearly dependent error vectors over $\mathbb{F}_{q^m}$ makes the linearization attack possible again. Recall the expression for the ciphertext:

$$\boldsymbol{r} = \boldsymbol{m}\mathbf{G} + \sum_{i=1}^{\ell} \alpha_i \boldsymbol{\rho}^{(i)} + \boldsymbol{e} = (\boldsymbol{m} + \sum_{i=1}^{\ell} \alpha_i \boldsymbol{m}'^{(i)})\mathbf{G} + \sum_{i=1}^{\ell} \alpha_i \boldsymbol{e}'^{(i)} + \boldsymbol{e}. \tag{22}$$

Let the rank of the error vectors $\boldsymbol{e}'^{(i)}$'s over $\mathbb{F}_{q^m}$ be denoted by u, i.e., u is the left-hand-side of (21). For an attacker, it is possible to identify u of the $\boldsymbol{\rho}^{(i)}$'s such that their corresponding errors $\boldsymbol{e}'^{(i)}$ are linearly independent. This can be done, for instance, by finding the $k+u$ linearly independent rows in the matrix $\begin{bmatrix} \mathbf{G} \\ \boldsymbol{\rho}^{(1)} \\ \vdots \\ \boldsymbol{\rho}^{(\ell)} \end{bmatrix}$. Without loss of generality, suppose these u vectors are $\boldsymbol{\rho}^{(1)}, \dots, \boldsymbol{\rho}^{(u)}$. Then, we can rewrite the ciphertext as

$$\boldsymbol{r} = (\boldsymbol{m} + \tilde{\boldsymbol{m}})\mathbf{G} + \sum_{i=1}^{u} \alpha_i' \boldsymbol{\rho}^{(i)} + \boldsymbol{e} \tag{23}$$

for unknowns $\alpha_i' \in \mathbb{F}_{q^m}$ and $\tilde{\boldsymbol{m}} \in \mathbb{F}_{q^m}^k$ is some unknown linear combination of the $\boldsymbol{m}'^{(i)}$ in the public key. The linearization attack will now succeed with high probability if $n \geq (u+2)w + k' + u$.[7] Note this is just (the negation of) condition (12) with ℓ replaced with u. Since u is much smaller, the attack succeeds in recovering $\boldsymbol{m} + \tilde{\boldsymbol{m}}$. The vector $\tilde{\boldsymbol{m}}$ is some linear combination of the vectors $\boldsymbol{m}'^{(i)}$ and since LIGA uses the shorter message variation, it is possible to unambiguously extract $\boldsymbol{m}$ from $\boldsymbol{m} + \tilde{\boldsymbol{m}}$.

6 Conclusion

To conclude, we will discuss some code families that do not seem to yield a secure instantiation of the framework and share some ideas for future research.

[7] k' was defined in Sect. 5.3 and equals $k - \ell$ since LIGA employs the shorter message variant. In fact, in LIGA it is necessary to use this variant since it is not possible to unambiguously extract the α_i's from $\boldsymbol{e} + \sum_{i=1}^{\ell} \alpha_i \boldsymbol{e}'^{(i)}$ during decryption since the $\boldsymbol{e}'^{(i)}$'s are linearly dependent over $\mathbb{F}_{q^m}$.

6.1 Insecure Code Classes

Reed-Solomon and Gabidulin Codes. As discussed in Sect. 5.3, both these code classes are insecure in our framework. This is because the linearization attack shows that the corresponding supercode decoding problem is easy when the parameters are chosen such that decryption is possible.

BCH and Other Structured Code Classes. Code classes that have a decoding procedure that involves solving a linear system of equations involving the syndrome coefficients (obtained from a "key equation") are likely to be insecure. This is because the framework can equivalently be formulated in a *parity-check version* (see Appendix C) where the public key and ciphertext are given by

$$\begin{aligned} \boldsymbol{\rho} &= \boldsymbol{e}'\mathbf{H}^\top, \\ \boldsymbol{y} &= \boldsymbol{e}\mathbf{H}^\top + \alpha\boldsymbol{\rho}. \end{aligned}$$

This implies that we can solve the supercode decoding problem by replacing each syndrome coefficient appearing in the linear system of the decoder with a term of the form $y_i - \alpha\rho_i$. Thus, we have only introduced one extra unknown α, and solving the decoding problem will be easy as long as we have enough equations. The latter is almost certainly true since the parameters need to be chosen such that decoding w' erasures and w errors simultaneously is possible, which implies that w cannot be its maximum value. Thus, many structured code classes, including BCH codes, which admit such a decoding procedure are insecure choices in the framework.

Low-Rank Parity Check (LRPC) Codes. LRPC codes [8] are another class of rank metric codes that have an efficient decoder [1] and, in fact, also an error-erasure decoder [10]. The concept of erasures used here corresponds to knowing the column support. We considered these codes in our framework, but the scheme turned out to be insecure. In further detail, we set up the public key and the ciphertext in the parity-check formulation (Appendix C) as

$$\begin{aligned} \boldsymbol{\rho} &= \boldsymbol{e}'\mathbf{H}^\top, \\ \boldsymbol{y} &= \boldsymbol{e}\mathbf{H}^\top + \alpha\boldsymbol{\rho}, \end{aligned}$$

where α is drawn from $\mathbb{F}_q^*$ instead of $\mathbb{F}_{q^m}^*$ to preserve the column support of $\boldsymbol{e}'$ and thus allow Alice to error-erasure decode (using the basic decoding algorithm described in, for example, [1]). Such a scheme can then be attacked as follows: compute the subspace $\mathcal{J} = \langle \boldsymbol{y}\rangle_{\mathbb{F}_q} + \langle \boldsymbol{\rho}\rangle_{\mathbb{F}_q}$ and the subspace $\mathcal{X} = \cap_i (f_i^{-1}\mathcal{J})$, where $(f_1, \ldots, f_d)$ is a basis of the d-dimensional $\mathbb{F}_q$-space spanned by the entries of the low-rank party check matrix $\mathbf{H}$ and $a\mathcal{J}$ denotes the subspace $\{ax \,|\, x \in \mathcal{J}\}$, where $a \in \mathbb{F}_{q^m}$. Experimentally, it turns out that $\mathcal{X}$ equals the space $\langle \boldsymbol{e}\rangle_{\mathbb{F}_q}$ with high probability. It is then straightforward to obtain the vector $\boldsymbol{e}$ by solving a linear system of equations.

6.2 Outlook

If a secure choice of code is possible then, given the above results, we think it is likely that the code family will not have much algebraic structure and would be more combinatorial in nature (LRPC codes notwithstanding). Of course, the code family needs to also have an error-erasure decoder, but for many code families, error-erasure decoding is possible as long as error decoding is possible.

Given the successful attacks outlined above, it is also plausible that no secure code class exists. Nonetheless, our framework is one way to realize an alternative PKE scheme where the secret information to decrypt is a property of the error rather than the code. The idea is natural, and even if not secure in its current form, investigating variations of the framework may lead to secure realizations. For instance, the current method of randomizing the public key by multiplying it with a non-zero field element may be overly simplistic and modifications in this direction are worth exploring.

Acknowledgments. This work has been supported by the ANR-DFG project "CROWD: Cryptography with Skew Codes", project number: WA3907/9-1. V. Weger is supported by the European Union's Horizon 2020 research and innovation programme under the Marie Skłodowska-Curie (Grant agreement No. 899987). This work is also partially supported by the Bavarian State Ministry of Science and Arts via the project EQAP.

A Hard Problem of the Interleaved Variant

The below statement applies to the Hamming metric but the rank-metric version can be easily obtained by replacing the condition $|\cup_{i=1}^{\ell} \operatorname{supp}(\boldsymbol{e}'^{(i)})| = w'$ with $|\operatorname{supp}(\boldsymbol{e}'^{(1)}) + \cdots + \operatorname{supp}(\boldsymbol{e}'^{(\ell)}))| = w'$ and with the understanding that $\operatorname{supp}(\cdot)$ now denotes the row support.

Problem 2 (Supercode Decoding with Interleaving $(\mathbf{G}, w, w', \ell)$*).* Draw uniformly at random $\boldsymbol{c}'^{(1)}, \ldots, \boldsymbol{c}'^{(\ell)} \in \langle \mathbf{G} \rangle$ and $\boldsymbol{e}'^{(1)}, \ldots, \boldsymbol{e}'^{(\ell)} \in \mathbb{F}_{q^m}^n$ such that

$$|\cup_{i=1}^{\ell} \operatorname{supp}(\boldsymbol{e}'^{(i)})| = w'.$$

Set $\boldsymbol{\rho}^{(i)} = \boldsymbol{c}'^{(i)} + \boldsymbol{e}'^{(i)}$ for $i = 1, \ldots, \ell$. Draw $\boldsymbol{c}^{\dagger} \in \left\langle \begin{bmatrix} \mathbf{G} \\ \boldsymbol{\rho}^{(1)} \\ \vdots \\ \boldsymbol{\rho}^{(\ell)} \end{bmatrix} \right\rangle \setminus \langle \mathbf{G} \rangle$ and $\boldsymbol{e} \in \mathcal{S}_w$ uniformly at random. Set $\boldsymbol{r} = \boldsymbol{c}^{\dagger} + \boldsymbol{e}$. Given only $\mathbf{G}$, $\boldsymbol{\rho}^{(1)}, \ldots, \boldsymbol{\rho}^{(\ell)}$ and $\boldsymbol{r}$, find $\boldsymbol{e}$.

B Incompatible Inequalities

Let us recall the three required inequalities on the parameters, which are given by

$$2w + w' \leq n - k, \tag{11}$$
$$n < (\ell + 2)w + k' + \ell, \tag{12}$$
$$w' > \frac{\ell}{\ell + 1}(n - k). \tag{13}$$

The first two inequalities are equivalent, respectively, to

$$w \leq \frac{1}{2}(n - k - w'),$$
$$w > \frac{n - \ell - k'}{\ell + 2}.$$

Together these imply

$$\frac{n - \ell - k'}{\ell + 2} < \frac{1}{2}(n - k - w').$$

and rearranging shows this is equivalent to

$$w' < (n - k)\left(\frac{\ell}{\ell + 2}\right) + 2\left(\frac{\ell + k' - k}{\ell + 2}\right).$$

The first term on the right-hand side is less than the right hand-side of (13) and the second term is at most 2. Thus, for any parameter set that guarantees meaningful security, this condition contradicts (13).

C Parity-Check Version of Framework

As in the McEliece PKE, there is an equivalent "Niederreiter formulation" of the system which uses the parity-check matrix $\mathbf{H}$ of the public code instead (see Algorithms 7 and 8 below).

Algorithm 7: Keygen (Parity-Check Version)

Input: —
Output: public key $\boldsymbol{\rho} \in \mathbb{F}_{q^m}^{n-k}$, secret key $\boldsymbol{\kappa} \in \mathbb{F}_{q^m}^{n}$

```
e' ←$ F_{q^m}^n s.t. wt(e') = w'
ρ ← e'H^T
κ ← e'
return (ρ, κ)
```

$\boldsymbol{e}' \xleftarrow{\$} \mathbb{F}_{q^m}^n$ s.t. $\mathrm{wt}(\boldsymbol{e}') = w'$
$\boldsymbol{\rho} \leftarrow \boldsymbol{e}'\mathbf{H}^T$
$\boldsymbol{\kappa} \leftarrow \boldsymbol{e}'$
return $(\boldsymbol{\rho}, \boldsymbol{\kappa})$

Algorithm 8: Encryption (Parity-Check Version)

Input: message $\boldsymbol{e} \in \mathbb{F}_{q^m}^n$ of weight w, public key $\boldsymbol{\rho} \in \mathbb{F}_{q^m}^{n-k}$
Output: ciphertext $\boldsymbol{y} \in \mathbb{F}_{q^m}^{n-k}$

$\alpha \xleftarrow{\$} \mathbb{F}_{q^m}^*$
$\boldsymbol{y} \leftarrow \boldsymbol{e}\mathbf{H}^T + \alpha\boldsymbol{\rho}$
return $\boldsymbol{y}$

References

1. Aragon, N., Gaborit, P., Hauteville, A., Ruatta, O., Zémor, G.: Low rank parity check codes: new decoding algorithms and applications to cryptography. IEEE Trans. Inf. Theory **65**(12), 7697–7717 (2019). https://doi.org/10.1109/TIT.2019.2933535
2. Augot, D., Finiasz, M.: A public key encryption scheme based on the polynomial reconstruction problem. In: Biham, E. (ed.) Advances in Cryptology - EUROCRYPT 2003, pp. 229–240. Lecture Notes in Computer Science, Springer (2003). https://doi.org/10.1007/3-540-39200-9_14
3. Augot, D., Finiasz, M., Loidreau, P.: Using the Trace Operator to repair the Polynomial Reconstruction based Cryptosystem presented at Eurocrypt 2003 (2003). https://eprint.iacr.org/2003/209
4. Bleichenbacher, D., Kiayias, A., Yung, M.: Decoding of interleaved reed solomon codes over noisy data. In: Baeten, J.C.M., Lenstra, J.K., Parrow, J., Woeginger, G.J. (eds.) Automata, Languages and Programming, pp. 97–108. Lecture Notes in Computer Science, Springer (2003). https://doi.org/10.1007/3-540-45061-0_9
5. Bombar, M., Couvreur, A.: Decoding supercodes of Gabidulin codes and applications to cryptanalysis. In: Post-Quantum Cryptography: 12th International Workshop, PQCrypto 2021, Daejeon, South Korea, July 20-22, 2021, Proceedings, p. 3-22. Springer-Verlag, Berlin, Heidelberg (2021). https://doi.org/10.1007/978-3-030-81293-5_1
6. Coron, J.S.: Cryptanalysis of a public-key encryption scheme based on the polynomial reconstruction problem. In: Bao, F., Deng, R., Zhou, J. (eds.) Public Key Cryptography - PKC 2004, pp. 14–27. Springer (2004). https://doi.org/10.1007/978-3-540-24632-9_2
7. Faure, C., Loidreau, P.: A new public-key cryptosystem based on the problem of reconstructing p–polynomials. In: Ytrehus, Ø. (ed.) Coding and Cryptography, pp. 304–315. Lecture Notes in Computer Science, Springer (2006). https://doi.org/10.1007/11779360_24
8. Gaborit, P., Murat, G., Ruatta, O., Zemor, G.: Low rank parity check codes and their application to cryptography. In: Budaghyan, L., Helleseth, T., Parker, M.G. (eds.) The International Workshop on Coding and Cryptography (WCC 13), p. 13 p. (2013). https://hal.science/hal-00913719
9. Gaborit, P., Otmani, A., Kalachi, H.T.: Polynomial-time key recovery attack on the Faure–Loidreau scheme based on Gabidulin codes. Designs, Codes Cryptograph. **86**(7), 1391–1403 (2018). https://doi.org/10.1007/s10623-017-0402-0
10. Gaborit, P., Ruatta, O., Schrek, J., Zémor, G.: RankSign: an efficient signature algorithm based on the rank metric. In: Mosca, M. (ed.) Post-Quantum Cryptography, pp. 88–107. Springer International Publishing, Cham (2014). https://doi.org/10.1007/978-3-319-11659-4_6
11. Kiayias, A., Yung, M.: Cryptanalyzing the polynomial-reconstruction based public-key system under optimal parameter choice. In: Lee, P.J. (ed.) Advances in Cryptology - ASIACRYPT 2004, pp. 401–416. Lecture Notes in Computer Science, Springer (2004). https://doi.org/10.1007/978-3-540-30539-2_28
12. Lavauzelle, J., Loidreau, P., Pham, B.D.: RAMESSES, a rank metric encryption scheme with short keys (2019). https://doi.org/10.48550/arXiv.1911.13119
13. Loidreau, P.: Decoding rank errors beyond the error-correcting capability. In: ACCT 2010, Tenth International Workshop on Algebraic and Combinatorial Coding Theory (2006)

14. Ore, O.: On a Special Class of Polynomials. Trans. American Math. Society **35**(3), 559–584 (1933). https://doi.org/10.2307/1989849
15. Overbeck, R.: Public key cryptography based on coding theory. Dissertation, Technische Universität Darmstadt (2007). https://tuprints.ulb.tu-darmstadt.de/823/
16. Randriambololona, H.: The syzygy distinguisher. Cryptology ePrint Archive, Paper 2024/1193 (2024). https://eprint.iacr.org/2024/1193
17. Renner, J., Puchinger, S., Wachter-Zeh, A.: LIGA: a cryptosystem based on the hardness of rank-metric list and interleaved decoding. Designs, Codes Cryptogr. **89**(6), 1279–1319 (2021). https://doi.org/10.1007/s10623-021-00861-z
18. Schmidt, G., Sidorenko, V.R., Bossert, M.: Collaborative decoding of interleaved reed-solomon codes and concatenated code designs. IEEE Trans. Inf. Theory **55**(7), 2991–3012 (2009). https://doi.org/10.1109/TIT.2009.2021308
19. Sidorenko, V., Jiang, L., Bossert, M.: Skew-feedback shift-register synthesis and decoding interleaved gabidulin codes. IEEE Trans. Inf. Theory **57**(2), 621–632 (2011). https://doi.org/10.1109/TIT.2010.2096032

McEliece Parameter Sets Optimized for Processing in Memory Architectures

Cyrius Nugier[1(✉)], Jean-Christophe Deneuville[1], and Eleonora Guerrini[2]

[1] Fédération ENAC ISAE-SUPAERO ONERA, Université de Toulouse, Toulouse, France
{cyrius.nugier,jean-christophe.deneuville}@enac.fr
[2] LIRMM, Montpellier, France
eleonora.guerrini@lirmm.fr

Abstract. Parameter sets for post-quantum primitives have to be designed with their whole lifecycle in mind. Computers will likely evolve to include larger computing spaces, for instance, with processing in memory. This architecture change is already known to improve cryptographic performance. In this paper, we propose new parameter sets for Classic McEliece to benefit further from this change.

Keywords: Cryptography · Processing in Memory · Optimization

1 Introduction

Code-based cryptography finds its origins within McEliece's blueprint back in 1978 [16]. He suggested using a disguised binary Goppa code as the public key, keeping the original decoder as a secret. In 1986, Niederreiter [18] suggested two improvements: first, to use the dual approach, instantiated with parity-check matrices instead of generator matrices (yielding the so-called equivalent Syndrome Decoding problem); and second, to use generalized Reed-Solomon codes instead of binary Goppa codes. Several more constructions instantiating McEliece with different families of codes have subsequently been suggested. Unfortunately, instantiating McEliece with any family of codes other than binary Goppa codes has proved to be unsecure. In 2017, the Classic McEliece cryptosystem [5] was submitted to the NIST Post-Quantum Cryptography standardization process [19] (it is still being evaluated in the fourth round alongside two other code-based cryptosystems: BIKE [4] and HQC [2,3]). It has been designed with the following three principles in mind: keeping binary Goppa codes, since they have not shown weaknesses since 1978; using Niederreiter's dual approach, as it performs better in the regime of parameters considered for cryptography; and upscaling the parameters to meet modern security requirements (at least a 128-bit security).

C. Nugier—This work was supported by ICO, Institut Cybersécurité Occitanie, funded by Région Occitanie, France.

V. Weger et al. (Eds.): CBCrypto 2024, LNCS 15531, pp. 45–66, 2025.
https://doi.org/10.1007/978-3-031-90229-1_3

Classic McEliece has some advantages compared to other code-based primitives in the NIST standardization process. First, the cryptosystem has been recognized for its robustness. Assuming one relates Classic McEliece to McEliece, it has the best longevity record of the whole NIST standardization process. It also features the shortest ciphertexts of the whole competition (roughly λ bytes for λ bits of security). Classic McEliece also has one of the best encryption performance among the remaining candidates in the fourth round. All of these gives an advantage in terms of computation and bandwidth costs over time.

However, this cryptosystem is held back by the size of its public key and the time it takes to generate. The public key is a large binary matrix, that is reduced to its systematic form to reduce its size. This reduction is done by Gaussian elimination, which is by far the most time-consuming step of the key generation. The parameter sets in the submission were designed to maximize security for a given key size.

For a fair comparison between primitives, the performances of the candidates are evaluated on fixed target architectures: some general purpose processors, an ARM micro-controller (Cortex M4) and hardware implementations of FPGA (Artix 7). The algorithms are being optimized for these architectures, but each primitive does not benefit from them in the same way: for instance, Classic McEliece features a high level of parallelism, particularly in Gaussian elimination, which is not exploitable on these architectures.

Hopefully, a new kind of computers are being developed, with Processing In Memory (PIM) capabilities. In these architectures, some memory spaces like caches or RAM are being redesigned, allowing to perform some computation on operands of a much larger size than those in a processor's ALU, while having a minimal additional cost in die surface. For instance, a technology called bit-line computing allows to modify SRAMs by using the horizontal word line as activators, and vertical bit line as sensors. When one or two lines are activated, a current is produced, and new cells called column peripherals can compute binary operations over them and output the result in a destination-enabled line. The robustness of this process is reliable and has high industry standards [24]. These architectures are generic and significantly improve data management, multimedia applications, and convolutional neural networks.

In addition, it has been shown that Classic McEliece benefits strongly from these architectures, since the Gaussian elimination can be parallelized. For example in [22], in a small cache of 16 KB with a parallelism of 4096 bits, this algorithm got a 4× acceleration. On bigger caches, the speedup went up to 46×, or 19× when compared with AVX-like vectorization.

However, the current parameter sets for Classic McEliece limit the impact of these architectures: since data need to be aligned in memory for bit-line computation, padding spaces need to be added, and the full parallelism is not exploited. In addition, a new `Syndrome` algorithm is available in these architectures with different complexity parameters [22].

In this paper, we introduce new parameter sets for Classic McEliece, with security levels comparable or better than their NIST equivalents, while displaying better properties for utilization on PIM computers and sometimes also with better performances on current architectures.

We study two families of parameter sets: one that aims to reduce the main drawback of Classic McEliece, the key generation time, and another one that aims to improve its strong characteristic, the encryption time. We do not search to optimize decryption times because no parameter set change would make it positively enter into consideration when choosing Classic McEliece over another post-quantum primitive.

The paper starts with a presentation of the McEliece cryptosystem and its current parameter sets. Section 3 continues with a reminder of the modifications to Classic McEliece required for optimization on PIM architectures and their effects. In Sect. 4 we present the rationale for the optimization, the parameter sets we obtained, and their experimental validation. Finally, Sect. 5 contains discussion on the advantages and disadvantages of these new parameter sets and where bottlenecks remain.

2 Background

2.1 Codes

In Information Theory, error correcting codes are a tool for guaranteeing the integrity of a message that has been potentially damaged. Basically, a message of dimension k is encoded into a message of length n, where $n \geq k$, so that t errors can be corrected. The parameters n, k, t are linked: on one hand, the greater the difference $n - k$, the more errors can be corrected. On the other hand, it is interesting to keep $\frac{k}{n}$ as close to 1 as possible to not waste bandwidth. Many classical results bound the choice of these parameters. The need to combine codes with good parameters and efficient decoding techniques has led research into the construction of algebraic codes, such as Reed Solomon codes and Goppa codes. Algebraic constructions of codes have received a large interest beyond the scope of error correction (such as McEliece's construction) which plays a fundamental tool in cryptographic protocols.

2.2 Classic McEliece

The Classic McEliece cryptosystem has two main parameters, n the length of the code and t its error correction capacity. From (n, t) one can deduce $m = \lceil \log_2(n) \rceil$ and $k = n - mt$ the dimension of the code. The Classic McEliece Algorithms can be broadly decomposed into subroutines as follows:

- KeyGen:
  - `SKgen`: Generates the first part of the Secret Key by generating a irreducible polynomial $g \in \mathbb{F}_{2^m}[x]$. The second part of the matrix is a permutation of n elements $\alpha_1 \cdots \alpha_n$ of $\mathbb{F}_{2^m}$.
  - `PKgen`:

* `Init`: Permutes elements of $\mathbb{F}_{2^m}$ into L according to the permutation of the secret key.
* `Root`: Apply the polynomial g to each element of L.
* `Inv`: Calculate the inverse $g(L)$ and store it as inv.
* `Zero`: Set the entire memory space of the matrix mat to 0.
* `Fill`: Fills mat by iteratively doing t times a transposition of inv into rows of mat and a multiplication of inv by L.
* `Gaussian`: Reduces mat to its systematic form by Gaussian Elimination.
* `Copy`: Outputs the Public Key which is the non-identity part of the matrix.

- `Controlbits`: Changes the representation of the permutation in the secret key to improve decryption times. The algorithm uses a Benes network to describe which pairs of elements to swap to reorder the list.

– `Encrypt`:
  - `Gen_e`: Generates an error vector e in dense representation filled with t ones at random positions and zeroes everywhere else.
  - `Syndrome`: Multiplies the public key matrix by e.

– `Decrypt`: Uses secret key permutation to compute L, then uses the Berlekamb-Massey algorithm to check if the roots of the minimal polynomial correspond to the positions of the ones in the syndrome.

2.3 Classic McEliece Parameter Sets

The parameter sets specified in the NIST submission's design rationale document are designed to maximize security for given key sizes. In this paper, we name parameter sets after their optimization goal (for instance, S for security), their claimed security category (1, 3 or 5 for I, III, or V), and when useful a letter to distinguish multiple sets. For instance, NIST parameter sets are:

– **S1** (3488, 64) has optimal security when $PKsize < 2^{18}\,bytes$, and n, t are multiples of 32.
– **S3** (4608, 96) has optimal security when $PKsize < 2^{19}\,bytes$, and n, t are multiples of 32.
– **S5a** (6688, 128) has optimal security when $PKsize < 2^{20}\,bytes$, and n, t are multiples of 32.
– **S5b** (6960, 119) has optimal security when $PKsize < 2^{20}\,bytes$.
– **S5c** (8192, 128) has n and t powers of 2.

These parameter sets are also meant to be out of reach of some specific attacks :

– Distinguishers on high rate codes such as [13] works whenever $t < 20$ when $n \leq 8192$.
– Attacks on low rate codes such as [7] work whenever rates are below 0.3 .
– Attacks on low m values such as [8] work whenever $m = 2$.

2.4 Processing In Memory Bit-Line Computing

Improvements to the classical Von Neumann computer architecture are increasingly constrained by the time required to move data between memory and computational elements. To address this issue, one can either add memory into the processor or integrate computing power into data storage elements.

Processing in Memory (PIM) repurposes existing storage elements into computational ones with minimal increase in chip surface. This approach reduces data movement, increases parallelism, and results in significant savings in time and energy. The value of this concept has been repeatedly proven with multiple architectures, including single and multicore processors, and whether processing is performed in cache, near the cache, or in main memory. Experimental results consistently demonstrate gains in performance and energy savings, particularly in highly parallel applications [6,11,14,23].

For instance, the "bit-line computing" technology modifies SRAMs by adding vertical sensors over cells. This enables bit-line sensors to produce the results of binary operations and to and store them directly. The Compute Cache architecture [1] exemplifies this adaptation by allowing computation within the cache for only an 8% increase in cache area. This increases the potential level of parallelism far beyond vector acceleration, with operands reaching 4096 bits in an L1 cache and up to 8KB in an L3 cache. A computing cache enables all basic binary operations, copying and comparing, and has been proven to improve data management in some identified applications such as bitmap operations or convolutional neural networks. The results show a 70–90% reduction in energy consumption. [1].

However, the use of PIM architectures creates memory management challenges, as bit-line operations require precise alignment of operands in cache. A bank partition (BP) is a part of the cache in which all cache lines (typically 64 bytes) can be subject to cache operations. A cache controller interprets new instructions from the processor and manages the operations in the cache, ensuring data integrity. Since the same operation can be undertaken in each BP at the same time, in a cache with 8 BP, we therefore reach a 512 byte (4096 bit) parallelism.

These constraints on how parallel operations are conducted force the developer to control exactly how the variables are placed in memory, and padding is often used to ensure data alignment in cache.

3 Memory Allocation and Algorithms Adaptation

Existing work [22], shows that architectures including bit-line computing, even on small L1 caches (4096-bit parallelism), can significantly improve the performance of classic McEliece. Notably, the Gaussian elimination that was the bottleneck of the key generation is improved. The results in Table 1 are given by simulation of a 32-bit RISC-V core [17] with and without a computing cache.

However, these improvements require fine management of the memory allocation of the data. For instance, to perform the masking and xoring operations

Table 1. Cycle counts of Classic McEliece **S1** with cache computing, and `Syndrome` with the matrix in transposed form.

Architecture	`Gaussian`	`PKgen`	KeyGen	Encrypt	Tr. `Synd.`
RISC-V	793 189 167	840 551 389	1 316 632 015	3 012 219	796 970
RISC-V + PIM	17 098 761	377 406 084	542 756 873	734 918	47 311

between rows of the Public Key matrix contained in the Gaussian elimination, these rows have to be cache-aligned. This means padding each matrix row so that its size is a multiple of the cache parallelism. Table 2 shows that according to the parameter set, the padding can either be almost half the length of the row or not be required at all.

Table 2. Padding required for cache alignment, 8-BP, 64-byte lines (4096bit parallelism)

Param. Set	**S1**	**S3**	**S5a**	**S5b**	**S5c**
(n, t)	(3488, 64)	(4608, 96)	(6688, 128)	(6960, 119)	(8192, 128)
Cache operations	1	2	2	2	2
Padding	608 (15%)	3584 (44%)	1504 (18%)	1232 (15%)	0 (0%)

Depending on cache associativity, this makes cache misses increasingly probable and slow (more data needs to move to the cache, less rows fit simultaneously in cache). To remedy this issue, the most used variables (the mask and an auxiliary row) have a dedicated mapping in cache, and padding rows are added so that no matrix row will overwrite them in cache. As a result of this change and the increase in parallelism, the height of the matrix $m \times t$ is a much more important factor in PIM architectures than in classical ones.

As a consequence of the improvements to the Gaussian elimination, the bottleneck of key generation becomes `Controlbits`, a routine that changes the list representation of the secret key into successive permutations of the elements' space. This is made for the decryption to be fast, constant-time and resistant to timing attacks. However, this seems to not benefit as well from the PIM parallelism, notably due to the alignment constraints of bit-line computing.

Nevertheless, in [22] it is shown that the improvements provided by PIM architectures are creating new design spaces. For example, storing the public key in transposed form (which is done in negligible time compared to the `Controlbits` routine) allows a significantly faster encryption algorithm (see Table 1). More precisely, the `Syndrome` subroutine can be modified to benefit from more parallelism. An overview of the `Syndrome` variations is presented in Fig. 1.

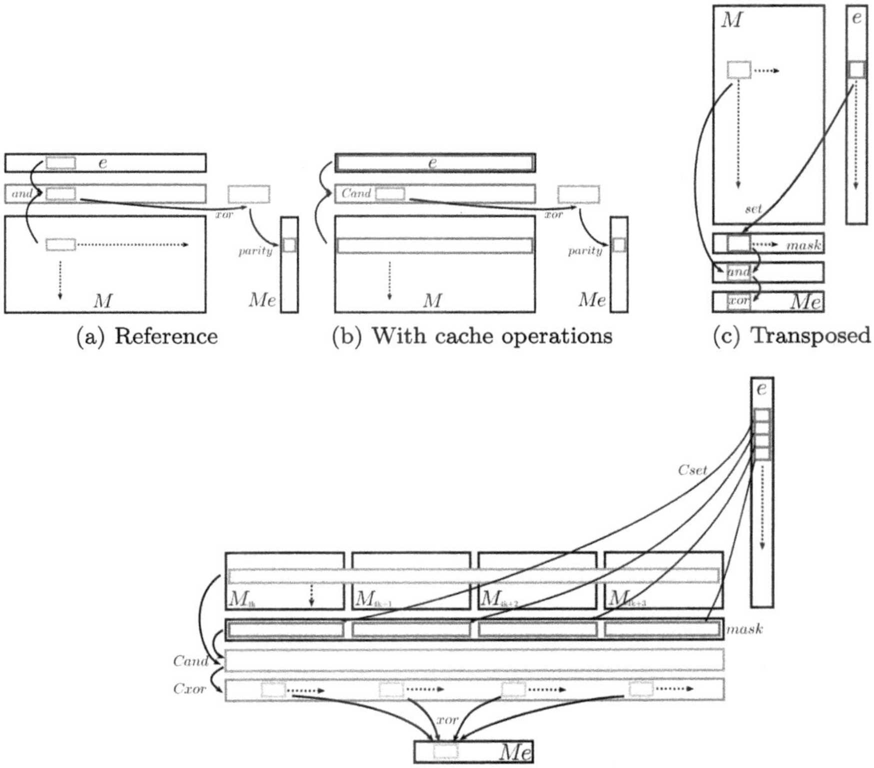

(a) Reference (b) With cache operations (c) Transposed

(d) Transposed with cache operations

Reference	Cache	Transposed	Transposed + Cache
$n \cdot m \cdot t/op$ **and**	$m \cdot t$ **Cand**	n **set**	n **Cset**
$n \cdot m \cdot t/op$ **xor**	$n \cdot m \cdot t/op$ **xor**	$n \cdot m \cdot t/op$ **and**	n/sub **Cand**
$m \cdot t$ **parity**	$m \cdot t$ **parity**	$n \cdot m \cdot t/op$ **xor**	n/sub **Cxor**
			$sub{\cdot}m{\cdot}t/op$ **xor**

(e) Operations for `Syndrome` variations. *op* is the operand size for classical operations, *sub* is the number of subdivisions of the transposed matrix.

Fig. 1. The Encrypt `Syndrome` subroutine variations on PIM architectures

The performance gain comes from multiple advantages of having the PK transposed, such as removing the parity calculation, and some loops over $m \cdot t$ as displayed in Fig. 1e. The number of subdivisions of the transposed PK matrix is limited by the number of BPs, since only a number of BPs that is a power of 2 can be used simultaneously for a cache operation. For example, with the parameter set **S1** (3488, 64), the row size of the transposed matrix is $64 \times 12 = 768$. As a BP is 512 bits long, padding is used to have rows of 1024 bits (2 BPs). Thus, in a 8-BP cache, we can have up to 4 subdivisions of the matrix (as represented in Fig. 1d). With larger parameters, it could be less, and with larger caches it could be more. Table 3 shows the padding required for NIST parameter sets.

Table 3. Padding required for cache alignment in transposed `Syndrome`, 8-BP, 64-byte lines (4096bit parallelism)

Param. Set	**S1**	**S3**	**S5a**	**S5b**	**S5c**
(n, t)	(3488, 64)	(4608, 96)	(6688, 128)	(6960, 119)	(8192, 128)
$m \cdot t$	768	1248	1664	1547	1664
BPs per subdivision	2	4	4	4	4
Padding	256 (25%)	800 (39%)	384 (18%)	501 (24%)	384 (18%)

Modifications to the Classic McEliece algorithms were made in order to use the parallelism of PIM architectures. While this alone gives significant performance improvement as shown in [22], this is far from optimal since a lot of the available parallelism is lost on the padding required for cache alignment (see Tables 2 and 3). In addition, these modifications drastically changed the way the parameter sets impact the complexity.

The goal of this paper is to push the optimization further, by finding alternate parameter sets that are equivalent or more secure than the NIST one while providing better performances.

4 New Parameter Sets, Impact on Performance

In the remainder of the paper, all security levels for parameter sets are evaluated with Crypto-TII estimators [9] and rounded down. It happens that for the range of parameter sets studied, May-Ozerov [15] is always the algorithm with the best time complexity, but not the best space complexity. The security of the NIST parameter sets can be found in Table 6.

Additionally, for the studied range of parameters, the rate is always above 0.48 [7] and lower than the critical rate [13], and m stays between 11 and 14 [8]. Also, whenever possible, we tried to keep the highest possible power of 2 in the factorization of n and t.

4.1 Designing Parameter Sets for KeyGen Time

In order to optimize the KeyGen algorithm, a few possible approaches can be taken:

- Improve the Gaussian elimination, by reducing the height on the matrix, since the complexity depends on $(m \cdot t)^2$. n may increase, but this can impact the number of cache operations for a single row of the matrix $C = \lceil n/cache\_parallelism \rceil$. Multiple performance sweetspots may be found when n is equal or just below a multiple of the cache parallelism.

- Improve the Gaussian elimination, by reducing the number of cache operations needed for a row of the matrix. This means reducing n down to a lower multiple of the cache parallelism, and raise t to compensate for the security.
- As shown in [22], the **`Controlbits`** subroutine is a new bottleneck of the key generation. As this algorithm only depends on $m = \lceil \log_2(n) \rceil$, we can improve the performance by reducing n to a lower power of 2, and raise t to compensate.

Therefore, we evaluated the security of various parameter sets with values of n fixed to powers of 2 or multiples of the cache parallelism. Then for each security category, we found the parameter set with the lowest $C \cdot (m \cdot t)^2$ value. This evaluation can be found in Table 4.

Table 4. Security of various parameter sets according to [9]. In bold our proposed parameter sets for KeyGen optimization.

t \ n	2048	4096	8192	12288	16384
160		180			
144		182			
128		182	275		
112		**180**	**259**		
96	103	174	251	275	
80	107	165	240	247	270
64	107	152	**194**	216	233
48	100	**133**	164	180	193
32		108	127	137	146
m	11	12	13	14	14

Legend
$256 \leq \lambda$
$192 \leq \lambda < 256$
$128 \leq \lambda < 192$

We noticed that a parameter set (4096, 112) has more security than **S3** (4608, 96), and a lower m, which is interesting for improving **`Controlbits`** performance. Therefore we decided to study it as a corresponding parameter set.

To recap, the parameter sets we introduce for KeyGen optimization are:

- **K1** (4069, 48) has optimal **`Gaussian`** complexity for $128 \leq \lambda$
- **K3a** (4096, 112) has optimal **`Controlbits`** complexity and better security than **S3** (4608, 96).
- **K3b** (8192, 64) has optimal **`Gaussian`** complexity for $192 \leq \lambda$.
- **K5** (8192, 112) has optimal **`Gaussian`** complexity for $256 \leq \lambda$.

Note that while **K5** (8192, 112) has better security than **S5a** (6688, 128) and **S5b** (6960, 119), it has lower security than **S5c** (8192,128) since it is strictly smaller, and **S5c** is over-dimentionned to powers of 2 and large security margins.

4.2 Designing Parameter Sets for Encryption Time with Transposed Syndrome and Cache Operations

As the complexity of the transposed `Syndrome` algorithm depends on the cache geometry (number of BPs), the chosen number of subdivisions (sub) and the operand size for non-aligned operations (op), it is less straightforward to optimize. The only impact parameter sets have on these is in the number of subdivisions. Cache operations are done in a number of BPs that is a powers of 2. We can therefore try to find how to minimize padding for various power of 2, i.e. for a fixed subdivision length l_{sub} and m find the largest t such that $m \cdot t$ remains lower than l_{sub}. From these m and t, we try to minimize n, but keep it as $m = \lceil \log_2(n) \rceil$. We evaluated the security for various pairs l_{sub}, m. Then for each security category we found the parameter set with the lowest n value. This evaluation can be found in Table 5.

Table 5. Security of various parameter sets according to [9]. In bold our proposed parameter sets for `Syndrome` optimization.

l_{sub}	t, m \ n	2560	3072	3584	4096	5120	6144	7168	8192	10240	12288	14336	16384
2048	170, 12		113	147	177								
2048	157, 13					209	244	**273**	297				
1024	85, 12	121	**140**	155	168								
1024	78, 13					177	**193**	206	217				
1024	73, 14									221	234	245	254
512	42, 12	104	112	118	124								
512	39, 13					126	**133**	139	144				

We noticed that the parameter set (3072, 85) is not directly comparable to (6144, 39) as a subdivision length two times smaller gives two times more subdivisions in a cache and maybe better performances. Therefore, we decided to study it as corresponding parameter sets.

In addition, for $t = 157$, we noticed a large security gap between $n = 6144$ and $n = 7168$. In order not to have an oversize n, we decided to use the middle value $n = 6656$, which has a security of $\lambda = 259$.

To recap, the parameter sets we introduce for Encrypt optimization are:

- **E1a** (3072, 85) has the smallest n for $l_{sub} = 512$ and $128 \leq \lambda$
- **E1b** (6144, 39) has the smallest n for $l_{sub} = 1024$ and $128 \leq \lambda$
- **E3** (6144, 78) has the smallest n for $l_{sub} = 1024$ and $192 \leq \lambda$
- **E5** (6656, 157) has the smallest n for $l_{sub} = 2048$ and $256 \leq \lambda$

Note: It is possible to choose to have $m > \lceil \log_2(n) \rceil$ in a parameter set, but this leads to lower security, or larger key sizes and lower performances for the same security.

4.3 Experiments

We ran our experiments on the open RISC-V simulator with cache computation [17] that was used in [22].

First, we compared the characteristics of the parameter sets (key and ciphertext sizes). Then, we compiled the cycle counts data:

- For each of the 13 parameter sets.
- For KeyGen and each subroutine (`SKgen`, `Init`, `Root`, `Inv`, `Zero`, `Fill`, `Gaussian`, `Copy`, `Controlbits`).
- For Encrypt, the `Gen_e` subroutine and all variations of `Syndrome` (Reference, Cache, Transposed, Transposed with cache).
  - For the transposed `Syndrome` with cache, multiple values of l_{sub}.
- For the Decrypt routine.
- With and without cache operations.
- For multiple cache configurations :
  - 8-BP, 32-way, representing a small 16kB cache similar to a L1
  - 8-BP, 256-way, representing a big cache similar to a L2 / L3
  - 8-BP, 4096-way. This is an ideal oversize cache that reduces data movement and shows changes in pure arithmetic speed.
  - 16-BP, 2048-way, 32-BP, 1024-way, 64-BP, 512-way. These are the same size as the previous one but with different geometries.

All algorithms that loop according to a random condition are computed for a single loop. We will study the probabilities associated with our parameter sets to account for this and deduce average cycle counts.

This data allows us to proceed to multiple studies and answer the following questions :

- How do the parameter sets we introduced compare to the NIST ones in the same security category.
- How performance is distributed among subroutines and where bottlenecks are. This will show where the targets for future optimizations are.
- How the presence of cache operations impacts performance.
- How the number of subdivisions impacts the performance of `Syndrome`.
- How the size of the cache impacts performance. This will tell how much arithmetic operations reductions are compensated with data movement increases.

4.4 Results

The volume of data collected is too large to be displayed here. All data is made available on [21] to allow for a more thorough comparison. Only essential data required to understand the main points of analysis is presented here. In addition, the software used to obtain the results is available at [20].

Characteristics. First, the change in parameter sets affects key sizes and ciphertext sizes. These are displayed in Table 6.

Table 6. Characteristics of the parameter sets. Key sizes in 10^3 bytes, ciphertext sizes in bytes. Light gray and dark gray are for notable decreases or increases respectively.

	Category I				Category III				Category V				
	S1	**K1**	**E1a**	**E1b**	**S3**	**K3a**	**K3b**	**E3**	**S5a**	**S5b**	**S5c**	**K5**	**E5**
λ	140	133	140	133	179	180	194	193	246	245	275	259	259
n	3488	4096	3072	6144	4608	4096	8192	6144	6688	6960	8192	8192	6656
t	64	48	85	39	96	112	64	78	128	119	128	112	157
m	12	12	12	13	13	12	13	13	13	13	13	13	13
PK_{size}	261	253	262	357	524	462	765	651	1045	1047	1358	1226	1178
SK_{size}	6.45	6.50	6.44	13.65	13.57	6.62	13.95	13.72	13.89	13.91	14.08	14.05	13.95
CT_{size}	96	72	128	64	156	168	104	127	208	194	208	182	256

Loops Depending on Random Variables. The two algorithms that significantly change their probability of success between parameter sets are the irreducible polynomial generation `genpoly` in `SKgen` and the generation of the noise indexes `indexes` in `Gen_e`. These changes are presented in Table 7. Other random algorithms keep the same success probabilities. The field ordering algorithm keeps a probability above 99% since $m \leq 13$ for all parameter sets, and the Gaussian elimination fail probability is the chance to generate a non-systematic matrix, which depends only on the field characteristic (2) and converges to 71.13% for any size above 20.

Table 7. Failure chances of various algorithms, in percentages. Light gray and dark gray are for notable decreases or increases respectively.

	Category I				Category III				Category V				
	S1	**K1**	**E1a**	**E1b**	**S3**	**K3a**	**K3b**	**E3**	**S5a**	**S5b**	**S5c**	**K5**	**E5**
`genpoly`	1.56	2.08	1.17	2.56	1.04	0.89	1.56	1.28	0.78	0.84	0.78	0.89	0.63
`indexes`	44.1	24.1	69.0	11.4	63.1	78.4	21.9	38.8	70.6	63.8	63.1	53.3	84.3

In regular architectures, these algorithms take a negligible time compared to other subroutines, so the impact on performance is small. However this stops being true on PIM architectures as the longer subroutines gets optimized. This is discussed more thoroughly in Sect. 5.

Performance. First, the performance on small regular architectures gives us an indication of the effect of a parameter set change on computers with low parallelism. The values found in Table 8.

Table 8. Performance of our parameter sets compared to NIST sets. Results in percentages of the baseline cycles, rounded to the nearest integer. Light gray and dark gray are for decreases or increases of more than 10% respectively. 8-BP, 32-way regular cache

	Category I			Category III			Category V					
	K1	**E1a**	**E1b**	**K3a**	**K3b**	**E3**	**K5**			**E5**		
Baseline	**S1**			**S3**			**S5a**	**S5b**	**S5c**	**S5a**	**S5b**	**S5c**
KeyGen	83	141	145	97	103	95	120	49	78	143	58	93
Encrypt	94	113	133	91	138	131	158	117	123	123	92	96
Decrypt	89	116	129	89	119	108	107	111	88	122	126	99

As parameter sets were designed to improve Keygen and Encrypt efficiencies in PIM architectures, improvements are expected to be found in these architectures. Figure 2 shows in detail the time taken by each subroutine.

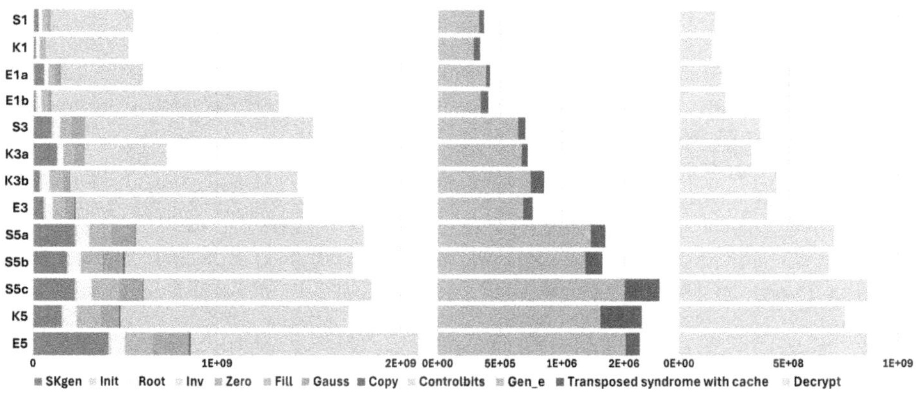

Fig. 2. Cycle counts for subroutines of KeyGen, Encrypt, and Decrypt. With cache operations, 8-BP, 4096-way.

5 Discussions and Perspectives

5.1 Parameter Sets Comparison

Characteristics

- For parameter sets with smaller security than their counterpart, e.g. **K1**, **E1b** compared to **S1** and **K5** compared to **S5c**, better performances are expected.
- On the contrary, for parameter sets with larger security e.g. **K3b**, **E3** compared to **S3**, and **K5** and **E5** compared to **S5a** and **S5b**, worse performances are expected.

- As the parameter sets introduced in this paper are not optimizing key sizes, contrary to NIST ones, they have larger keys, except **K1** and **K3a**.
- Two parameter sets have different m than their NIST counterpart, **K3a** has $m = 12$ while **S2** has $m = 13$ and thus the Secret Key is twice smaller. On the contrary, **Eb1** has $m = 13$ while **S1** has $m = 12$, and the secret key is twice larger. However, these variations will not be sufficiently influential decision criteria to choose or reject Classic McEliece rather than another primitive.

KeyGen

- For parameter sets optimizing KeyGen time, as t was often decreased and n increased, the ciphertext size $(m \cdot t)$ is often reduced, which makes this advantage of Classic McEliece even more advantageous.
- For, **K3a** the reduction in m reduces the size of the public key $(m \cdot t)(n - m \cdot t)$, reducing one of the downsides of the primitive.
- For parameter sets aimed at improving encryption, there is no common direction in the variation in n and t and therefore no noteworthy impact on key sizes.
- In every architecture the parameter sets designed for KeyGen performance have faster Key generation, with the only exception being **K5** compared to **S5a** on 32-way cache without PIM since there is more data movement that is not compensated by faster operations.
- `SKgen`, `Init`, `Root`, `Inv` and `Zero` follows variations similar to t^2, SK_{bytes}, n, $n \cdot SK_{bytes}$, and PK_{size} respectively. On caches with PIM, `Gaussian` time follows $C \cdot (m \cdot t)^2$ while on caches without PIM, it follows $n(m \cdot t)^2$ as expected.
- While the `Gaussian` step is faster in the **K** sets than in **S** sets whenever PIM is available, for the 32-way no PIM cache, **K3b** and **K5** have a 3% and 20% slower KeyGen than **S3** and **S5a** respectively because of this step.
- The `Copy` subroutine has two implementations, a very efficient one for whenever $tail = (m \cdot t) \, mod \, 8 = 0$, and a slower one for the other cases, which happens typically in the **E** parameter sets where $m = 13$ and are not designed to have t contain powers of 2 factors.
- This slowdown is due to the arbitrary choice of having padding bits at the end of each row and not at the start. While this design choice could be changed for improvements, this subroutine is negligible so it can be left as is.
- The `Controlbits` subroutine takes constant time in m, this is the reason why **E1b** KeyGen in very slow in each architecture, and why **K3a** is faster, even on architectures in which `Gaussian` is slower.

Encrypt

- The **E** parameter sets perform worse than expected for the whole Encrypt algorithm, they are in most cases even slower than their counterparts.
- The **E** parameter sets were designed to improve the `Syndrome` subroutine since it was the computational bottleneck of the Encrypt algorithm. However, it happens that when the `Syndrome` is in its transposed version with cache operations is becomes so fast compared to the `Gen_e` subroutine that the parameter changes impact negatively Encrypt.
- Hopefully, according to expectations, transposed `Syndrome` with cache operations is improved by **E** parameter sets on architectures with PIM, and the transposed version is improved in architectures without PIM, except for **E1b**, that has a taller matrix due to the increased m.
- In the non-transposed representation of the public key, the change from **S** parameter sets to any other except **K1** and **K3a** has a negative impact on `Syndrome` performance across architectures.
- Using t values that are not powers of 2 in the **E** parameter sets introduces a $tail = m \cdot t \, mod \, 8 \neq 0$, which causes slowdown. While this slowdown was negligible in KeyGen's `Copy` subroutine, it is significant in `Syndrome`.

Computational Bottlenecks

- The Gaussian elimination remains the main bottleneck on regular architectures (approx. 80% of KeyGen) and on PIM architectures (approx. 50%). This is due to data movement, because once it is removed (the cache is large enough) it takes less than 10% of the KeyGen.
- On the opposite, `Controlbits` was important only on parameters sets with low t and high m on regular architectures, but becomes predominant in PIM architectures.
- Among the remaining KeyGen subroutines, `Init`, `Root`, `Inv` and `Zero` remain negligible. However, once the `Gaussian` is optimized, `SKgen` can take up to 20% of KeyGen time (12% for **S** parameter sets).
- In order to allow for the faster transposed `Syndrome` algorithm (with or without PIM), the transposition should be integrated into KeyGen as a replacement of `Copy`. The naive bit-wise transposition we estimated takes between 1% and 12% depending on parameter set and architecture but this could be optimized in software and in hardware and is expected to be negligible after optimization.
- When transposed, `Syndrome` without cache operations takes around 80% of the Encrypt algorithm. Transposed and with cache operations, it takes between 20% and 5% when no data movement is involved, and around 40% in a 32-way cache. In these situations, `Gen_e` becomes a new bottleneck.

Impact of Cache Operations

- The main time gained in KeyGen comes from the Gaussian elimination. With cache operations in a 32-way cache, it takes 30% or less than with a regular cache, except for **S3** which was already identified as poorly adapted to PIM.
- With no data movement (the time for arithmetical operations) measured in the 4096-way cache represents just 2% or less than the original time, confirming the results of [22].
- Similar improvements are visible for the `Zero` step.
- The total KeyGen time gain varies a lot with parameter set and architecture but is regularly between 50% and 30% of the original time.
- For `Syndrome`, the transposed version with cache operations takes from 12% of the original time on a 32-way cache down to around 4% in the 4096-way cache.
- The non-transposed `Syndrome` with cache takes from 90% to 70% of original time, this is mainly because accumulation is still done the original way.
- The total Encrypt time with cache operation goes from 43% to 24% to the original time.

Impact of Subdivisions on the Transposed Syndrome with Cache Operations

- The behavior of the `Syndrome` time according to the number of BPs (and therefore of subdivisions) is approximately the same for all parameter sets, as can be seen on Fig. 3. As could be expected, increasing the number of PBs can improve the performances by reducing the height of the transposed matrix subdivisions (n/sub), but increasing it too much causes the final accumulation to take longer.
- The optimal number of subdivisions depends strongly on the machine used, notably in the time taken for the final accumulation. For instance, in a 64-bit RISCV, or a chip with non-aligned parallelism (e.g. AVX), we could expect the lowest point of the curve to be more on the right, meaning time will be gained by increasing even more the number of subdivisions.

5.2 Removing the `Gen_e` bottleneck with regularity

As mentioned before, for parallel architectures, the transposed `Syndrome` algorithm makes `Gen_e` the computational bottleneck. Optimizing parameters for `Syndrome` only can have a negative impact on KeyGen because of a slowdown in `Gen_e`. This routine can be decomposed into the following steps:

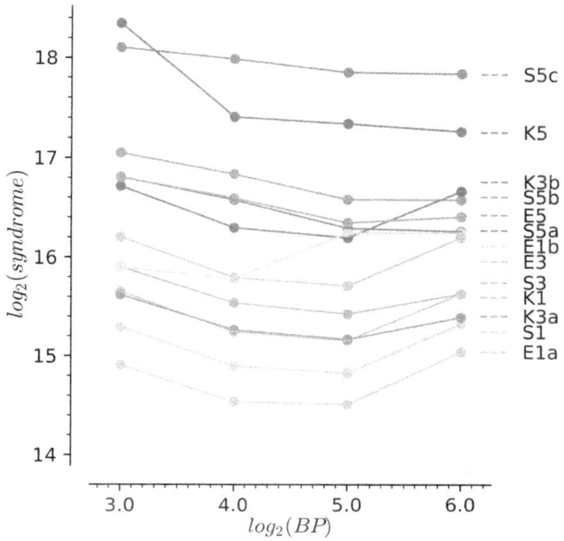

Fig. 3. `Syndrome` cycle counts for various cache geometries. Same cache size from 4096-way 8-BP to 512-way 64-PB.

- `Indexes`: generate the indexes of the t noise bits
  - `Rng` $\mathcal{O}(4t)$: for $2t$ possible indexes, generate $8 \cdot \left\lceil \frac{\lceil \log_2(n) \rceil}{8} \right\rceil$ random bits
  - `Apply_mask` $\mathcal{O}(4t)$: keep only $\lceil \log_2(n) \rceil$ bits per index
  - `Count_weight` $\mathcal{O}(t\frac{2^{\lceil \log_2(n) \rceil}}{n})$: keep the first t indexes that are smaller than n. Each specific index has a $\frac{n}{2^{\lceil \log_2(n) \rceil}}$ chance of being kept. Since $2t$ indexes were generated, there is a probability $P_{weight\_fail} = \sum_0^t \binom{2t}{i} \cdot p^i \cdot (1-p)^{2t-i}$ that the generation did not give enough good indexes and has to be restarted. On the parameter sets studied in this paper this probability is always below 1% except for **S3** (3.5%) since n is just over a power of 2 and t is not so big.
  - `Check_diff` $\mathcal{O}(t^2/2)$: check for each index that none of the following indexes are equal to it. Because of the birthday paradox this happens with probability $P_{same\_fail} = \frac{n!}{n^t \cdot (n-t)!}$. On the parameter sets studied in this paper this probability is between 12% and 85%.
- `Densify` $\mathcal{O}(n \cdot t)$: create a dense vector from indexes by looping over each index and the length of the vector (n) to place each 1 in constant time.

When considering all these algorithms alongside the `Syndrome` complexity, optimizing parameter sets for encryption is a lot harder and more dependent on implementation constants. However, in order to optimize `Gen_e` one could change how the noise vector is generated, and instead generate it by parts, as a *regular* vector.

Equivalently as presented in [12], a vector $\mathbf{e} \in \mathbb{F}_2^n$ of weight t is called *r-regular* if it can be written as $\mathbf{e} = (\mathbf{e}_1, \mathbf{e}_2, \cdots, \mathbf{e}_r)$, where each $\mathbf{e}_i$ has length n/r and Hamming weight t/r.

We extend this concept by allowing r not to divide the $gcd(n,t)$ by introducing pseudo-regularity. With $//$ and $\%$ being the Euclidean quotient and remainder operators, a vector $\mathbf{e} \in \mathbb{F}_2^n$ of weight t is called *r-pseudo-regular* if it can be written as $\mathbf{e} = (\mathbf{e}_1, \mathbf{e}_2, \cdots, \mathbf{e}_r)$, where the first $n\%r$ vectors have length $n//r+1$ and the rest length $n//r$, and the first $t\%r$ vectors have weight $t//r+1$ and the rest weight $t//r$.

The Syndrome Decoding Problem can be adapted to require only r-regular or r-pseudo-regular solutions. For instance, the Regular Syndrome Decoding (RSD) problem is defined and studied in [12] with $r = t$. A similar Pseudo Regular Syndrome Decoding (PRSD) problem can be defined for pseudo-regularity with an expected loss of hardness. Also, as increasing r reduces the space of possible $\mathbf{e}$ vectors, $r = t$ seems to be the situation with the worst security.

The estimator provided with [10] was used to compute the estimated security for the parameter sets studied in this paper. It showed that even if asymptotically the $1 - k/n$ and t/n of all parameter sets are in a zone in which RSD should be harder, this in not the case for most of them. The only parameter sets loosing security with t-regular vectors are **K3a** ($\lambda = 175$), **E1a** ($\lambda = 136$), and **E5** ($\lambda = 257$).

Regularity affects positively all the `Gen_e` steps:

- `Rng` now has to generate r times $8 \cdot \left\lceil \frac{\lceil \log_2(n/r) \rceil}{8} \right\rceil$ random bits. This means that if r is big enough, only half as much random bits are required. This is the case with $r \geq 16$ for parameter sets with $m = 12$ or $r \geq 32$ for parameter sets with $m = 13$.
- Similarly, `Apply_mask` time can be halved under the same conditions, or even removed if $\lceil \log_2(n/r) \rceil = 8$.
- `Count_weight` probability of failing increases slightly: the probability of each index being good remains the same (and is always >0.5) but the probability of finding t/r good indexes among $2t/r$ possible ones is lower as r increases. Hopefully, the cost of a reroll is divided by r so this makes on average a time reduction to reduce r.
- `Check_diff` time is divided by r since it compares t/r indexes in time $(t/r)^2/2)$ and has to be done r times. Additionally, the chances of failing are significantly reduced.
- Similarly, `Densify` is r times faster since it is done on r vectors, and each vector takes $(\frac{n}{r} \cdot \frac{t}{r})$.

Furthermore, when n/r is a power of 2, there is no need to do `Count_weight` and only t indexes must be generated by `Rng` and `Apply_mask`.

Although increasing r seems good, at the implementation level there is a drop in performance whenever n/r is no longer a multiple of 8 or whenever regularity has to be replaced with pseudo-regularity. This can be observed in Fig. 4. A slowdown can also be observed when r gets too big.

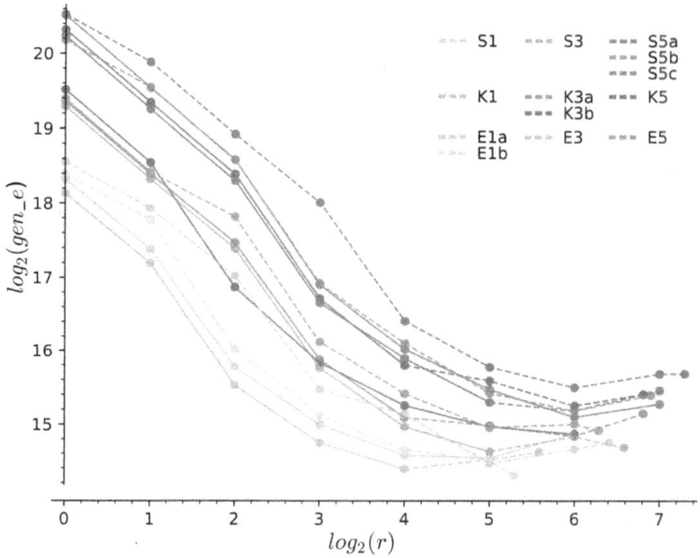

Fig. 4. `Gen_e` cycle counts for various levels of regularity (r). Log_2 scale. Dashed lines indicate pseudo-regularity.

Replacing `Gen_e` by its regular version with the best possible r reduces the Encrypt total time to approx. 20% with the transposed `syndrome` and cache operations, and to approximately 80% with the regular `syndrome` without cache operations. With large PIM architectures, the parameter sets **K3a**, **K3b**, **E1a**, and **E5** are better than their corresponding **S** set. This is not the case for **E1a** in small regular architectures.

5.3 Parameter Sets Recommendations

In order to make the following recommendations, we are considering that the choice of Classic McEliece is made over other cryptographic primitives because of the small ciphertext size, the fast encryption, and the confidence on the security assumptions. Additionally, this choice of primitive is not made in a context in which key size or decryption times are important. However, improvements in KeyGen times and adaptability to architecture changes are desired.

Category I: For most cases, **S1**(3488,64) seems to remain the best choice for $\lambda \geq 128$. However, **K1**(4096,48) could be considered as an alternate for specific situations as its ciphertexts are 25% shorter, and overall performance is better.

Category III: The current parameter set would benefit from being replaced by **K3a**(4096,112), since it has overall better performance now and in the future. However, to satisfy $\lambda \geq 192$, **K3b**(8192,64) would be the better replacement.

Category V: S5a (6688,128) and **S5b**(6960,119) are broadly equivalent. With PIM architectures and adapted algorithms (`Gen_e` regularity and `Syndrome` transposition), **S5a** will have a slightly faster Encrypt and **S5b** a slightly faster KeyGen. However, to satisfy $\lambda \geq 256$, **K5** (8192,112) is the best choice. In any case, **S5c**(8192,128) should be kept as a more secure alternate since its performance will improve a lot from more parallel architectures.

5.4 Perspectives

On an implementation level the future possible continuations of this work on Classic McEliece would be to produce definitive optimized software, and to use the non-aligned parallelism to study notably how fast the encryption could be in the best conditions. The remaining bottlenecks identified above can be alleviated: `SKgen` contains a Gaussian elimination that could also benefit from parallelism, and space efficiency could be improved by placing this matrix in the padding bits of the much bigger PK matrix. In addition, PK matrix filling can be done with PIM operations called bit-array operations. Finally, a `Controlbits` algorithm designed to consider parallelization would help greatly KeyGen performances.

To extend this work further, the same process of optimization, redesign, and parameter sets adaptation should be done with other post-quantum primitives such as BIKE and HQC so that the comparison of code-based primitives is fair.

6 Conclusion

In this paper, we have investigated the impact on performance of adapting Classic McEliece parameter sets for Processing In Memory (PIM) architectures. Through experimental analysis and comparison with regular architectures, we have proposed and studied 8 new parameter sets.

Our findings show that PIM opens design spaces for new algorithms. The transposed `Syndrome` algorithm is so efficient on PIM architectures that `Gen_e` would become a bottleneck, which is why we proposed to generate the vector as regular (by blocks), for a strong performance.

These algorithms and Keygen's parallelized `Gaussian` routine do not have the same complexity in the code parameters as the original ones. This difference encourages the adaptation of the current parameter sets.

In summary, our work demonstrates the impact of parameter set adaptation for Classic McEliece on PIM architectures. It enables new recommendations for parameter sets across different security categories.

References

1. Aga, S., Jeloka, S., Subramaniyan, A., Narayanasamy, S., Blaauw, D., Das, R.: Compute caches. In: 2017 IEEE International Symposium on High Performance Computer Architecture (HPCA), pp. 481–492. IEEE (2017)
2. Aguilar Melchor, C., et al.: Hamming Quasi-Cyclic (HQC) (Nov 2017). https://hal.science/hal-01946880, submission to the NIST post quantum standardization process (2017)
3. Aguilar-Melchor, C., Blazy, O., Deneuville, J.C., Gaborit, P., Zémor, G.: Efficient encryption from random quasi-cyclic codes. IEEE Trans. Inf. Theory **64**(5), 3927–3943 (2018). https://doi.org/10.1109/TIT.2018.2804444
4. Aragon, N., et al.: Bike: bit flipping key encapsulation (2022)
5. Bernstein, D., et al.: Classic McEliece (2017)
6. Boroumand, A., et al.: Lazypim: Efficient support for cache coherence in processing-in-memory architectures (2017)
7. Carrier, K., Debris-Alazard, T., Meyer-Hilfiger, C., Tillich, J.P.: Statistical decoding 2.0: Reducing decoding to lpn (2022). https://arxiv.org/abs/2208.02201
8. Couvreur, A., Otmani, A., Tillich, J.P.: Polynomial time attack on wild McEliece over quadratic extensions. Cryptology ePrint Archive, Paper 2014/112 (2014). https://eprint.iacr.org/2014/112
9. Crypto-TII: Cryptographicestimators. https://github.com/Crypto-TII/CryptographicEstimators (2022)
10. Crypto-TII: Cryptographicestimators. https://github.com/Memphisd/Regular-ISD (2023)
11. Draper, J., et al.: The architecture of the diva processing-in-memory chip (2002)
12. Esser, A., Santini, P.: Not just regular decoding: Asymptotics and improvements of regular syndrome decoding attacks. Cryptology ePrint Archive, Paper 2023/1568 (2023). https://eprint.iacr.org/2023/1568
13. Faugère, J.C., Gauthier, V., Otmani, A., Perret, L., Tillich, J.P.: A distinguisher for high rate McEliece cryptosystems. Cryptology ePrint Archive, Paper 2010/331 (2010). https://eprint.iacr.org/2010/331
14. Li, S., Niu, D., Malladi, K.T., Zheng, H., Brennan, B., Xie, Y.: Drisa: A dram-based reconfigurable in-situ accelerator (2017)
15. May, A., Ozerov, I.: On computing nearest neighbors with applications to decoding of binary linear codes. In: Oswald, E., Fischlin, M. (eds.) Advances in Cryptology - EUROCRYPT 2015, pp. 203–228. Springer, Berlin Heidelberg, Berlin, Heidelberg (2015)
16. McEliece, R.J.: A public-key cryptosystem based on algebraic coding theory. Coding Theory **4244**, 114–116 (1978)
17. Migliore, V.: Crypto riscv (2022). https://gitlab.laas.fr/vmiglior/crypto-riscv
18. Niederreiter, H.: Knapsack-type cryptosystems and algebraic coding theory. Prob. Contr. Inform. Theory **15**(2), 157–166 (1986)
19. NIST: Post-quantum cryptography (2016). https://www.nist.gov/pqcrypto
20. Nugier, C.: RISC-V PIM simulator with alternate McEliece algorithms (2024). https://archive.softwareheritage.org/browse/origin/?origin_url=https://gitpub.recherche.enac.fr/McEliece-ParamSets
21. Nugier, C., Deneuville, J.C.: McEliece parameter sets optimized for processing in memory architectures - additional material (2024). https://doi.org/10.57745/JL7JDM

22. Nugier, C., Migliore, V.: Acceleration of Classic McEliece post-quantum cryptosystem with cache processing. IEEE Micro (2023)
23. Wang, J., Wang, J., et al.: A 28-nm compute sram with bit-serial logic/arithmetic operations for programmable in-memory vector computing. IEEE J. Solid-State Circ. **55**(1), 76–86 (2019)
24. Wang, J., et al.: A 28-nm compute sram with bit-serial logic/arithmetic operations for programmable in-memory vector computing. IEEE J. Solid-State Circ. **55**(1), 76–86 (2020). https://doi.org/10.1109/JSSC.2019.2939682

Breaking HWQCS: A Code-Based Signature Scheme from High Weight QC-LDPC Codes

Alex Pellegrini[1] and Giovanni Tognolini[2](✉)

[1] Department of Mathematics, Eindhoven University of Technology, Eindhoven, The Netherlands
alex.pellegrini@live.com

[2] Department of Mathematics, University of Trento,Trento, Italy
giovanni.tognolini@unitn.it

Abstract. We analyse HWQCS, a code-based signature scheme presented at ICISC 2023, which uses quasi-cyclic low-density parity-check codes (QC-LDPC). The scheme introduces high Hamming weight errors and signs each message using a fresh ephemeral secret key rather than using only one secret key, so to avoid known attacks on QC-LDPC signature schemes. In this paper, we show that the signatures of HWQCS leak substantial information concerning the ephemeral keys and formally describe this behaviour. Furthermore, we show that for each security level, we can exploit the leakage to efficiently reconstruct partial secret data from very few signatures, and finally mount a universal forgery attack.

Keywords: post-quantum cryptography · code-based cryptography · hash&sign signature · universal forgery · information leakage

1 Introduction

Linear codes are well known in cryptography, and their use spans multiple branches of this subject. In this work we deal with a code based digital signature scheme, and here we provide an overview of how linear codes have been used with this aim. Two important computationally-hard algebraic problems arising in coding theory are the *Syndrome Decoding Problem* (SDP) and the *Codeword Finding Problem* (CFP). Several code-based digital signature schemes are built upon SDP, and an adversary who wants to break such schemes is usually left with the only option of solving these problems. Currently, the two main approaches for solving these problems are information set decoding (ISD) algorithms and the generalized birthday algorithm (GBA). ISD algorithms demonstrate greater efficiency when the decoding problem involves only a limited number of solutions, while GBA is more effective when the expected number of solutions is high. We refer the reader to [32] for more details on these techniques.

V. Weger et al. (Eds.): CBCrypto 2024, LNCS 15531, pp. 67–93, 2025.
https://doi.org/10.1007/978-3-031-90229-1_4

Code-based digital signature schemes can be divided into two variants. On one hand there are the schemes based on proofs of knowledge, which in turn can follow the Fiat-Shamir [12] or the Schnorr-Lyubashevsky [19] approach, while on the other side there are the schemes following the hash&sign paradigm.

According to the aim of this work, we focus on digital signature schemes which follow the hash&sign paradigm. Among these schemes, CFS [5,6] and KKS [16] were the first to be introduced. The idea behind both of them is the same: the digest of a message to be signed is considered as the syndrome of a corrupted codeword, and the owner of the private key is the only one capable of decoding. The main difference between CFS and KKS is that in the former the digest is considered as a random syndrome, leading to an inefficient scheme, while the latter manipulates the digest to output a decodable syndrome, leading to a very efficient but insecure scheme [22]. In order to mitigate the efficiency issues of CFS, some authors have proposed similar schemes in which the hash function is replaced by a map whose output is a syndrome with small-weight coset-leader, combining some ideas behind both CFS and KKS. This approach has been adopted e.g. in [27], where the authors use a hash function based on the works of Augot, Finiasz and Sendrier [2] and on the Merkle-Damgard construction [7,21]. However, this approach has been proved insecure in [11]. In 2014, the first rank metric code based signature scheme RankSign [14] has been designed, which uses augmented low-rank parity check codes (LRPC). A structural attack [9] on RankSign was developed in 2018. During the last years there have been different attempts to build similar digital signature schemes [3,13,14,17,18,23,25,28,30] but it seems that the cryptographic community is far from achieving a trustable proposal [1,4,10,15,24,26,29,33]. In this context, a remarkable hash&sign code based digital signature scheme that still remains unbroken is Wave [8], which has been proposed in 2019 and relies on the indistinguishability of the normalized generalized $(U, U+V)$ codes. In 2023, a new attempt to build a code based signature scheme, called HWQCS [31], has been made. HWQCS uses QC-LDPC codes with the Hamming metric and introduces the use of high weight errors to make the decoding problem harder for an attacker. The construction of this scheme is devised in such a way to prevent other known attacks from being effective.

Contributions. This work addresses the cryptanalysis of [31]. First, we show that every signature reveals some information about the ephemeral keys used in the signing process. We formally describe the stochastic behaviour of the exposed information leakage and provide experimental evidence in support of our claims. As a second contribution, we show that, for any security level (128, 192 and 256 bits) an attacker only needs a handful of signatures in order to be able to recover partial secret data, which is meant to be known only by the signer. We compute lower bounds of the success probability of this event for every intercepted signature. After, we proceed to mount a universal forgery attack using the data recovered using our strategy. Finally, we provide a probabilistic algorithm along with a SageMath implementation and supporting experimental data, that efficiently outputs the secret data needed to forge valid signatures for arbitrary messages. We also propose a slight speed-up for the security level 256 which turns out to be the best scenario for our attack.

Related Works. Part of this work follows a strategy similar to that of [29]. The idea is to analyse a valid signature and exploit the bias on the distribution of the signature coefficients to perform a statistical test and thus obtain part of the key.

Organization of this Paper. Section 2 defines the necessary background and notation which will be used throughout the paper. In Sect. 3 we recall the setup, key generation, sign and verification phases of the HWQCS scheme. In Sect. 4 we describe how the scheme leaks information and how to exploit very few signatures to completely reconstruct the private ephemeral values used to sign. Finally, Sect. 5 presents our attack in its full generality, showing how to mount a universal forgery attack, and we propose a speed-up for the security level 256. We provide Sagemath code and experimental data in support of our results.

2 Background and Notation

Let $\mathbb{F}_2$ be the finite field with two elements. The Hamming weight of a vector $\mathbf{a} \in \mathbb{F}_2^n$, with $\mathbf{a} = (a_1, a_2, \ldots, a_n)$, is defined as

$$\mathsf{wt}_H(\mathbf{a}) := \#\{i \in [n] \mid a_i \neq 0\},$$

where we denote with $[n]$ the interval of values $\{1, \ldots, n\}$. The Hamming distance between two vectors $\mathbf{a}, \mathbf{b} \in \mathbb{F}_2^n$ is defined as $d_H(\mathbf{a}, \mathbf{b}) = \mathsf{wt}_H(\mathbf{a} - \mathbf{b})$. We denote with $V_{n,w} \subset \mathbb{F}_2^n$ the set of vectors of length n and Hamming weight w. Let $k, n \in \mathbb{N}$, then a linear $[n, k, d]$-code C is a k-dimensional subspace of $\mathbb{F}_2^n$ with minimum distance

$$d := \min_{\mathbf{a},\mathbf{b} \in C, \mathbf{a} \neq \mathbf{b}} d_H(\mathbf{a} - \mathbf{b}).$$

In this work, we consider vectors as row vectors, and thus their transposes as column vectors. Let $R := \mathbb{F}_2[x]/(x^n - 1)$ be the ring of all polynomials over $\mathbb{F}_2$ of degree less than n. For $\mathbf{a} \in \mathbb{F}_2^n$, denote by $\mathbf{a}(x)$ the unique polynomial $\mathbf{a}(x) = a_1 + a_2 x + \cdots + a_n x^{n-1} \in R$. Given a vector $\mathbf{a} \in \mathbb{F}_2^n$, we denote its j-th entry by a_j when $\mathbf{a} = (a_1, \ldots, a_n)$ is specified, and by $(\mathbf{a})_j$ otherwise. Similarly, we denote the j-th coefficient of a polynomial $\mathbf{a}(x) \in R$ as $(\mathbf{a}(x))_j$. Consider two elements $\mathbf{a}, \mathbf{b} \in \mathbb{F}_2^n$, then the vector $\mathbf{c} \in \mathbb{F}_2^n$, such that $\mathbf{c}(x) = \mathbf{a}(x)\mathbf{b}(x)$, can be computed by the formula $\mathbf{c}^\top = \mathsf{circ}(\mathbf{a}) \cdot \mathbf{b}^\top$, where $\mathsf{circ}(\mathbf{a})$ is the $n \times n$ circulant matrix obtained from $\mathbf{a}$ as

$$\mathsf{circ}(\mathbf{a}) = \begin{pmatrix} a_1 & a_n & \cdots & a_2 \\ a_2 & a_1 & \cdots & a_3 \\ \vdots & \vdots & \ddots & \vdots \\ a_n & a_{n-1} & \cdots & a_1 \end{pmatrix},$$

whose j-th row is a circular right shift of the first row of j positions. Due to this relationship, it is possible to link the description of the scheme presented in [31] to the Syndrome Decoding Problem.In particular, HWQCS uses two ephemeral values $\mathbf{e}_1(x), \mathbf{e}_2(x) \in R$ with low weight, a public element $\mathbf{h}(x) \in R$, and computes the public value $\mathbf{b}(x) = \mathbf{e}_1(x) \cdot \mathbf{h}(x) + \mathbf{e}_2(x) \cdot \mathbf{h}^{-1}(x)$. An attacker who wants to retrieve the ephemeral values is therefore asked to solve the decoding problem given by $\mathbf{b} = (\mathsf{circ}(\mathbf{h}) \,|\, \mathsf{circ}(\mathbf{h})^{-1}) (\mathbf{e}_1, \mathbf{e}_2)^\top$.

3 The HWQCS Signature Scheme

This section describes the algorithms of HWQCS [31] for key generation KeyGen, signature Sign and the verification Verify. We assume that the global parameters $k, w_f, w_u, w_e, w_c, w_s, w_t, R$ and hash_{ω_c} provided in the setup are public, and we do not specify them as input of the algorithms.

Setup. The parameters of the scheme are the positive integers $k, w_f, w_u, w_e, w_c, w_s, w_t$, the quotient ring $R = \mathbb{F}_2[x]/(x^k - 1)$ and hash_{ω_c} a hash function with fixed output Hamming weight w_c

Key Generation. The Key Generation algorithm KeyGen works as follows. First, two elements $\mathbf{f}_1$ and $\mathbf{f}_2$ are randomly chosen from the set V_{k,w_f}. These vectors must satisfy the condition that their associated polynomials $\mathbf{f}_1(x)$ and $\mathbf{f}_2(x)$ are invertible in R. Then, the public key pk is the product of the inverse of $\mathbf{f}_1(x)$ by $\mathbf{f}_2(x)$. Finally, the private key sk is the pair $(\mathbf{f}_1, \mathbf{f}_2)$. Algorithm 3.1 describes the key generation process for HWQCS.

Algorithm 3.1. Key generation algorithm KeyGen

Input : $\emptyset$.
Output : a public key $\mathsf{pk} \in R$ and a private key $\mathsf{sk} \in R^2$.

1. Choose random $\mathbf{f}_1, \mathbf{f}_2 \in V_{k,w_f}$ such that $\mathbf{f}_1(x), \mathbf{f}_2(x) \in R$ are invertible.
2. Compute $\mathbf{h}(x) := \mathbf{f}_1(x)^{-1} \cdot \mathbf{f}_2(x) \in R$.
3. Output the public key $\mathsf{pk} = \mathbf{h}$ and the private key $\mathsf{sk} = (\mathbf{f}_1, \mathbf{f}_2)$.

Signature Algorithm. The signing algorithm Sign takes as inputs a message m and the public and secret keys $(\mathsf{pk}, \mathsf{sk})$, and works as follows. It starts by randomly selecting vectors $\mathbf{e}_1$ and $\mathbf{e}_2$ from the set V_{k,w_e}, as well as $\mathbf{u}_1$ and $\mathbf{u}_2$ from V_{k,w_u}. Next, two auxiliary polynomials are computed. The first is $\mathbf{b}(x)$, which depends on $\mathbf{h}(x), \mathbf{e}_1(x)$ and $\mathbf{e}_2(x)$. The second is $\mathbf{d}(x)$, derived from the private key components $\mathbf{f}_1(x)$ and $\mathbf{f}_2(x)$ along with $\mathbf{u}_1(x)$ and $\mathbf{u}_2(x)$. A hash function hash_{ω_c} is then applied to the message m and the values $\mathbf{b}$, $\mathbf{d}$, and $\mathbf{h}$, producing a vector $\mathbf{c}$ in V_{k,w_c}. After, the values $\mathbf{s}_1$ and $\mathbf{s}_2$ are computed as

$\mathbf{s}_i(x) := \mathbf{u}_i(x)\mathbf{f}_i(x)+\mathbf{c}(x)\mathbf{e}_i(x)$, for $i = 1, 2$. Notice that, according to the parameters that will be chosen later, each $\mathbf{s}_i$ is obtained by multiplying and summing sparse polynomials and, moreover, one of such polynomials is public. This will be the starting point for our attack. The algorithm then verifies that the weights of $\mathbf{s}_1$, $\mathbf{s}_2$, and $\mathbf{d}$ meet the required constraints not exceeding some threshold parameters w_s and w_t. If any of these weights exceed such limitations, the algorithm restarts. The signature σ consists of the elements $(\mathbf{c}, \mathbf{b}, \mathbf{s}_1, \mathbf{s}_2)$. Algorithm 3.2 describes the process of generating a signature σ for a message m using a public key pk and a private key sk.

Algorithm 3.2. Signature algorithm Sign

Input : the message to sign m, a public key $\mathsf{pk} = \mathbf{h} \in \mathbb{F}_2^n$ and a secret key $\mathsf{sk} = (\mathbf{f}_1, \mathbf{f}_2) \in \mathbb{F}_2^n \times \mathbb{F}_2^n$.
Output : a valid signature $\sigma \in (\mathbb{F}_2^n)^4$ for m under pk and sk.

1. Choose random $\mathbf{e}_1, \mathbf{e}_2 \in V_{k,w_e}$ and $\mathbf{u}_1, \mathbf{u}_2 \in V_{k,w_u}$.
2. Set
   - $\mathbf{b}(x) := \mathbf{e}_1(x)\mathbf{h}(x) + \mathbf{e}_2(x)\mathbf{h}^{-1}(x) \in R$ and
   - $\mathbf{d}(x) = \mathbf{u}_1(x)\mathbf{f}_2(x) + \mathbf{u}_2(x)\mathbf{f}_1(x)$.
3. Compute $\mathbf{c} := \mathsf{hash}_{\omega_c}(m, \mathbf{b}, \mathbf{d}, \mathbf{h}) \in V_{k,w_c}$.
4. Compute $\mathbf{s}_i(x) := \mathbf{u}_i(x)\mathbf{f}_i(x) + \mathbf{c}(x)\mathbf{e}_i(x)$ for $i = 1, 2$.
5. If $w(\mathbf{s}_1) > w_s$ or $w(\mathbf{s}_2) > w_s$ or $w(\mathbf{d}) > w_t$ then repeat from Step 1.
6. With this notation, the signature is given by $\sigma = (\mathbf{c}, \mathbf{b}, \mathbf{s}_1, \mathbf{s}_2)$.

Verification Algorithm. The verification algorithm begins by computing the polynomial $\mathbf{t}(x)$ using the the signature $(\mathbf{c}, \mathbf{b}, \mathbf{s}_1, \mathbf{s}_2)$ and the public key pk. Specifically, $\mathbf{t}(x)$ is computed as $\mathbf{s}_1(x)\mathbf{h}(x) + \mathbf{s}_2(x)\mathbf{h}^{-1}(x) - \mathbf{c}(x)\mathbf{b}(x)$. Next, a hash digest $\mathbf{c}' \in V_{k,w_c}$ is computed applying hash_{ω_c} to the message m and the values $\mathbf{b}$, $\mathbf{t}$, and $\mathbf{h}$. The algorithm then checks if $\mathbf{c} = \mathbf{c}'$, the weight $\mathsf{wt}_H(\mathbf{t})$ is less than or equal to w_t, and $\mathbf{t}(x) \neq 0$. If all conditions are satisfied, the signature is accepted as valid; otherwise, it is rejected. Algorithm 3.3 describes the process of verifying a signature σ for a message m using a public key pk.

Algorithm 3.3. Verification algorithm Verify

Input : A message m, a public key $\mathsf{pk} = \mathbf{h} \in \mathbb{F}_2^n$ and a signature $\sigma = (\mathbf{c}, \mathbf{b}, \mathbf{s}_1, \mathbf{s}_2) \in (\mathbb{F}_2^n)^4$.
Output : accept or reject signature σ for m.

1. Compute $\mathbf{t}(x) := \mathbf{s}_1(x)\mathbf{h}(x) + \mathbf{s}_2(x)\mathbf{h}^{-1}(x) - \mathbf{c}(x)\mathbf{b}(x)$.
2. Compute $\mathbf{c}' := \mathsf{hash}_{\omega_c}(m, \mathbf{b}, \mathbf{t}, \mathbf{h}) \in V_{k,w_c}$.
3. If $\mathbf{c} = \mathbf{c}'$, $\mathsf{wt}_H(\mathbf{t}) \leq w_t$ and $\mathbf{t}(x) \neq 0$, then accept, otherwise reject.

The sets of parameters suggested for 128,192 and 256 bits of security are reported in Table 1.

Table 1. Suggested parameters for security levels 128,192 and 256 of HWQCS.

Security	k	ω_f	ω_u	ω_e	ω_c	ω_s	ω_t
128	12539	145	33	141	31	4844	4937
192	18917	185	41	177	39	7450	7592
256	25417	201	51	191	51	10111	10216

4 Information Leakage of HWQCS

We start this section by giving an overview of our attack, which serves as a motivation for the study of the objects developed in the rest of the section, where we show that the signatures generated by the Sign algorithm of HWQCS leak a critical amount of information about the ephemeral values $\mathbf{e}_1$ and $\mathbf{e}_2$.

4.1 Overview of the Attack

The idea is to use information set decoding techniques in a simplified scenario, which allows us to perform fast recovery of secret data, hence forge signatures. In other words, we aim at finding error free positions in the vector $\mathbf{e}$. To this end, we apply a similar analysis to [29], which manipulates each intercepted signature in such a way that it is possible to separate a good amount of zero bits of $\mathbf{e}_1$ and $\mathbf{e}_2$ from the one bits. Furthermore, we can carefully guess additional zero bits. Once k, zero bits have been recovered, we can perform linear algebra using a submatrix of the public matrix

$$\mathbf{H} = \left[\mathsf{circ}(\mathbf{h}) \mid \mathsf{circ}(\mathbf{h})^{-1}\right]. \tag{1}$$

to reconstruct the values $\mathbf{e}_1, \mathbf{e}_2$ completely. Finally, compute $\mathbf{u}_i(x)\mathbf{f}_i(x) = \mathbf{s}_i(x) - \mathbf{c}(x)\mathbf{e}_i(x)$ for $i = 1, 2$. At this point it is possible to start forging valid signatures, see Sect. 5 for a detailed explanation and Algorithm 5.1 for the concrete attack.

In the next section, we will formally describe the behaviour of the information leakage, which will be useful to determine the necessary objects for our attack. Our analysis is identical for both $i = 1, 2$, therefore we won't fix a value of i.

4.2 Statistical Analysis of the Information Leakage

We describe the technique introduced in [29] adapted to the setting of HWQCS. Let $\sigma = (\mathbf{c}, \mathbf{b}, \mathbf{s}_1, \mathbf{s}_2)$ be an intercepted signature where $\mathbf{s}_i(x) := \mathbf{u}_i(x)\mathbf{f}_i(x) + \mathbf{c}(x)\mathbf{e}_i(x)$, as per Sign. We can expose many one bits of $\mathbf{e}$ as follows:

– Let $v \in \mathsf{supp}(\mathbf{c})$ and write

$$\begin{aligned} x^{-v}\mathbf{s}_i(x) &= x^{-v}\left(\mathbf{u}_i(x)\mathbf{f}_i(x) + \mathbf{c}(x)\mathbf{e}_i(x)\right) \\ &= x^{-v}\mathbf{u}_i(x)\mathbf{f}_i(x) + \left(1 + x^{-v}\sum_{l\in\mathsf{supp}(\mathbf{c})\setminus\{v\}} x^l\right)\mathbf{e}_i(x) \\ &= \mathbf{e}_i(x) + x^{-v}\mathbf{u}_i(x)\mathbf{f}_i(x) + \sum_{l\in\mathsf{supp}(\mathbf{c})\setminus\{v\}} x^{l-v}\mathbf{e}_i(x). \end{aligned} \tag{2}$$

– Finally, compute

$$\mathbf{d}_i(x) := \sum_{v\in\mathsf{supp}(\mathbf{c})} x^{-v}\mathbf{s}_i(x) \in \mathbb{Z}[x], \tag{3}$$

where the sum in 3 is taken over $\mathbb{Z}$. Notice that $x^{-v}\mathbf{u}_i(x)\mathbf{f}_i(x)$ is a polynomial whose coefficient vector is a circular left shift of the coefficient vector of $\mathbf{u}_i(x)\mathbf{f}_i(x)$ of v positions. The same holds for $x^{l-v}\mathbf{e}_i(x)$, so that for every $v \in \mathsf{supp}(\mathbf{c})$, the value $x^{-v}\mathbf{s}_i(x)$ is given by $\mathbf{e}_i(x)$ plus some random noise, see (2). Therefore, we expect the larger coefficients of $\mathbf{d}_i(x)$ to be associated with the entries of $\mathbf{e}_i(x)$ equal to 1. Figures 1, 2 and 3 give a visual representation of the information leakage for a random signature, of an instance of HWQCS for security levels 128, 192 and 256.

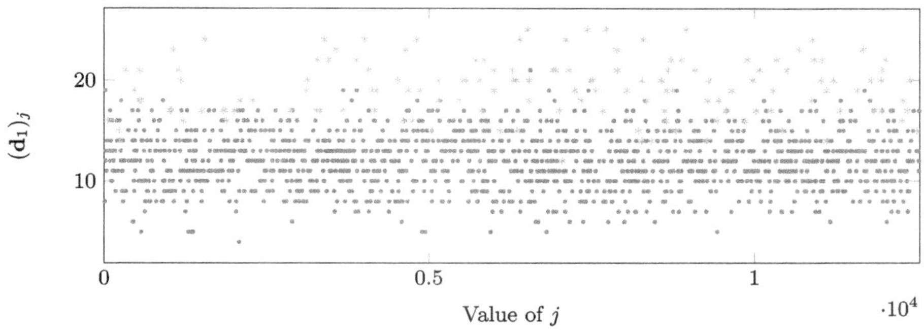

Fig. 1. Information leakage of $\mathbf{e}_1(x)$ of an instance of HWQCS with security parameter 128. We highlighted with blue dots the entries j for which $(\mathbf{e}_1)_j = 0$, and with orange asterisks the entries j for which $(\mathbf{e}_1)_j = 1$. (Color figure online)

From Figs. 1, 2 and 3, one can easily see that, for each security level, the one entries of $\mathbf{e}_i$ (in the examples in the figures $i = 1$) are pushed to the upper half of the plot, meaning that they correspond to the higher coefficients of $\mathbf{d}_i(x)$. This means that, with great probability, a random entry $(\mathbf{d}_i)_j \in \mathbb{Z}$ which is small enough will correspond to a zero entry $(\mathbf{e}_i)_j$. The rest of this section is dedicated to the proof of Theorem 1, which describes the behaviour of the entries of $\mathbf{d}_i \in \mathbb{Z}^k$.

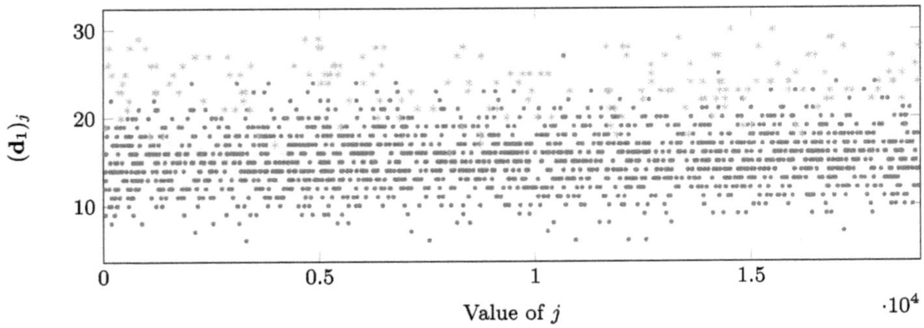

Fig. 2. Information leakage of $\mathbf{e}_1(x)$ of an instance of HWQCS with security parameter 192. We highlighted with blue dots the entries j for which $(\mathbf{e}_1)_j = 0$, and with orange asterisks the entries j for which $(\mathbf{e}_1)_j = 1$. (Color figure online)

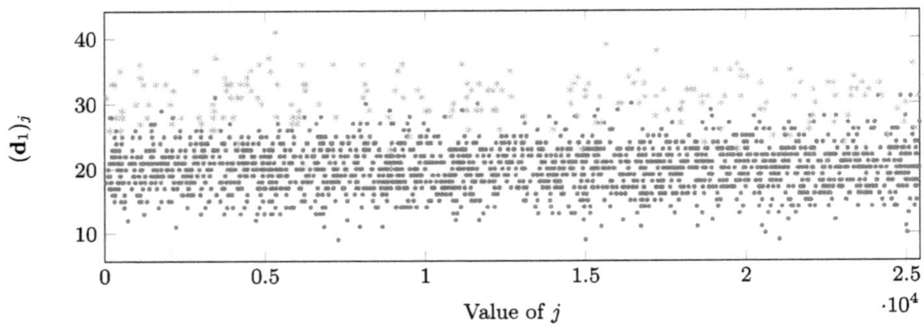

Fig. 3. Information leakage of $\mathbf{e}_1(x)$ of an instance of HWQCS with security parameter 256. We highlighted with blue dots the entries j for which $(\mathbf{e}_1)_j = 0$, and with orange asterisks the entries j for which $(\mathbf{e}_1)_j = 1$. (Color figure online)

Theorem 1. *Let $\mathbf{u}(x), \mathbf{f}(x), \mathbf{e}(x)$ and $\mathbf{c}(x)$ be random elements of R, in such a way that $\mathsf{wt}_H(\mathbf{u}) = w_u, \mathsf{wt}_H(\mathbf{f}) = w_f, \mathsf{wt}_H(\mathbf{e}) = w_e$ and $\mathsf{wt}_H(\mathbf{c}) = w_c$, respectively. Let $v \in \mathsf{supp}(\mathbf{c})$ and define*

$$\mathbf{d}(x) := \sum_{v \in \mathsf{supp}(\mathbf{c})} \left(\mathbf{e}_i(x) + x^{-v}\mathbf{u}_i(x)\mathbf{f}_i(x) + \sum_{l \in \mathsf{supp}(\mathbf{c}) \setminus \{v\}} x^{l-v}\mathbf{e}_i(x) \right) \in \mathbb{Z}[x].$$

Assume that the random variables $x^{-v}\mathbf{u}_i(x)\mathbf{f}_i(x)$ and $\sum_{l \in \mathsf{supp}(\mathbf{c}) \setminus \{v\}} x^{l-v}\mathbf{e}(x)$ are independently distributed, as v and l vary. Then, for $j \in [k-1]$, we have that $(\mathbf{d}(x))_j$ is binomially distributed as

$$(\mathbf{d}(x))_j \sim Bin\left(w_c, \prod_{i=1}^{3} p_i + \sum_{i=1}^{3} p_i \prod_{j \in [3], j \neq i} (1 - p_j) \right), \tag{4}$$

where

$$p_1 = w_e/k, \qquad p_2 = (1-(1-2w_e/k)^{w_c-1})/2$$

and

$$p_3 = \frac{1}{\binom{k}{w_u}\binom{k}{w_f}} \sum_{\substack{1\leq l\leq \min(w_u,w_f)\\ l \text{ odd}}} \binom{k}{l}\binom{k-l}{w_u-l}\binom{k-w_u}{w_f-l}.$$

Let $j \in [k-1]$. We will analyse the distribution of $(x^{-v}\mathbf{s}_i(x))_j$ for a fixed $v \in \mathsf{supp}(\mathbf{c})$, breaking down the analysis on the three summands of the last equation in (2), treated as random variables. Namely, we are going to compute the probability distributions of $(x^{-v}\mathbf{u}_i(x)\mathbf{f}_i(x))_j$ and of $\left(\sum_{l\in\mathsf{supp}(\mathbf{c})\setminus\{v\}} x^{l-v}\mathbf{e}_i(x)\right)_j$, while we consider $(\mathbf{e}_i)_j$ to be Bernoulli distributed with parameter ω_e/n. After, we will provide the explicit probability distribution of the sum of the three random variables and describe the distribution of $(\mathbf{d}_i(x))_j = \left(\sum_{v\in\mathsf{supp}(\mathbf{c})} x^{-v}\mathbf{s}_i(x)\right)_j$, where the sum is taken over $\mathbb{Z}$.

The following result comes in handy to compute the probability distribution of the entry $(x^{-v}\mathbf{u}_i(x)\mathbf{f}_i(x))_j$.

Lemma 1. *[20, Proposition 2.4.1] Let $\mathbf{u}(x), \mathbf{f}(x) \in R$ be random elements such that $\mathsf{wt}_H(\mathbf{u}) = \omega_u$ and $\mathsf{wt}_H(\mathbf{f}) = \omega_f$. Set $\mathbf{z}(x) = \mathbf{u}(x)\mathbf{f}(x)$, then for every $j \in \{0,\ldots,k-1\}$, we have that $(\mathbf{z}(x))_j$ is distributed as a Bernoulli random variable with parameter $p = P(z_j = 1)$ equal to:*

$$p = \frac{1}{\binom{k}{w_u}\binom{k}{w_f}} \sum_{\substack{1\leq l\leq min(w_u,w_f)\\ l \text{ odd}}} \binom{k}{l}\binom{k-l}{w_u-l}\binom{k-w_u}{w_f-l}. \tag{5}$$

Lemma 1 provides us with a way to compute the distribution of $(\mathbf{u}_i(x)\mathbf{f}_i(x))_j$.

Remark 1. Note that $x^{-v}\mathbf{u}_i(x)\mathbf{f}_i(x)$ is a polynomial whose coefficient vector is a circular left shift of the coefficient vector of $\mathbf{u}_i(x)\mathbf{f}_i(x)$ of v positions. Therefore, the two random variables $(\mathbf{u}_i(x)\mathbf{f}_i(x))_j$ and $(x^{-v}\mathbf{u}_i(x)\mathbf{f}_i(x))_j$ are identically distributed.

We move on to the distribution of $\sum_{l\in\mathsf{supp}(\mathbf{c})\setminus\{v\}} x^{l-v}\mathbf{e}_i(x)$. We will treat $(x^{l-v}\mathbf{e}_i(x))_j$ as independent random variables, as l and v vary.

Lemma 2. *Let $X_1, X_2, \ldots, X_k$ be k independent random variables following a Bernoulli distribution with parameter $q < \frac{1}{2}$ and let $X = \sum_{h=1}^{k} X_h$ be their sum over $\mathbb{F}_2$. Then X is Bernoulli distributed with parameter $p = \mathbb{P}(X = 1)$ equals to:*

$$p = \left(\frac{1}{2} - \frac{(1-2q)^k}{2}\right). \tag{6}$$

Proof. The random variable X is Bernoulli distributed with parameter recursively given by $T(k) = T(k-1)(1-q) + q(1-T(k-1))$, with $T(0) = 0$. Our goal is to convert this expression into a closed formula. Consider the formal power series $f(x) := \sum_{h=1}^{\infty} T(h)x^h$ where we consider the real interval $x \in (0,1)$. We have that

$$\begin{aligned}
f(x) &= \sum_{h=1}^{\infty} T(h)x^h \\
&= qx + \sum_{h=2}^{\infty} T(h)x^h \\
&= qx + \sum_{h=2}^{\infty} (T(h-1)(1-q) + q(1-T(h-1)))x^h \\
&= qx + (1-q)\sum_{h=2}^{\infty} T(h-1)x^h + \sum_{h=2}^{\infty} qx^h - q\sum_{h=2}^{\infty} T(h-1)x^h \\
&= \sum_{h=1}^{\infty} qx^h + (1-q)xf(x) - qxf(x) \\
&= \sum_{h=1}^{\infty} qx^h + (1-2q)xf(x).
\end{aligned}$$

We can rewrite $f(x)$ as

$$\begin{aligned}
f(x) &= \frac{\sum_{h=1}^{\infty} qx^h}{1-x+2qx} \\
&= \frac{q \cdot (\frac{x}{1-x})}{1-x+2qx} \\
&= \frac{qx}{(1-x)\cdot(1-(1-2q)x)} \\
&= \frac{\frac{1}{2}}{(1-x)} - \frac{\frac{1}{2}}{(1-(1-2q)x)}
\end{aligned}$$

For every constant A such that $|Ax| < 1$ we have that $\sum_{h=0}^{\infty}(Ax)^h = 1/(1-Ax)$, therefore we can rewrite both $1/(1-x)$ and $1/(1-(1-2q)x)$ as formal power series, obtaining

$$f(x) = \frac{1}{2}\sum_{h=0}^{\infty} x^h - \frac{1}{2}\sum_{h=0}^{\infty}(1-2q)^h x^h = \sum_{h=1}^{\infty}\left(\frac{1}{2} - \frac{(1-2q)^h}{2}\right)x^h.$$

We conclude that $T(h) = \frac{1}{2} - \frac{(1-2q)^h}{2}$, as claimed. □

We have all the necessary tools to be able to take a big step forward towards the proof of Theorem 1.

Proposition 1. *Let* $\mathbf{u}(x), \mathbf{f}(x), \mathbf{e}(x)$ *and* $\mathbf{c}(x)$ *be random elements of* R, *in such a way that* $\mathsf{wt}_H(\mathbf{u}) = w_u, \mathsf{wt}_H(\mathbf{f}) = w_f, \mathsf{wt}_H(\mathbf{e}) = w_e$ *and* $\mathsf{wt}_H(\mathbf{c}) = w_c$, *respectively. Let* $v \in \mathsf{supp}(\mathbf{c})$, *define* $\mathbf{s}(x) := \mathbf{u}(x)\mathbf{f}(x) + \mathbf{c}(x)\mathbf{e}(x)$ *and consider the expression*

$$x^{-v}\mathbf{s}(x) = \mathbf{e}_i(x) + x^{-v}\mathbf{u}_i(x)\mathbf{f}_i(x) + \sum_{l \in \mathsf{supp}(\mathbf{c}) \setminus \{v\}} x^{l-v}\mathbf{e}_i(x).$$

Assume that the random variables $\sum_{l \in \mathsf{supp}(\mathbf{c}) \setminus \{v\}} x^{l-v}\mathbf{e}(x)$ *are independently distributed, as* l *and* v *vary. Then, for* $j \in [k-1]$, *we have that* $(x^{-v}\mathbf{s}(x))_j$ *is distributed as a Bernoulli random variable with parameter* $p = \mathbb{P}((x^{-v}\mathbf{s}(x))_j = 1)$ *equal to:*

$$p = \prod_{i=1}^{3} p_i + \sum_{i=1}^{3} p_i \prod_{j \in [3], j \neq i} (1 - p_j), \tag{7}$$

where

$$p_1 = w_e/k, \qquad p_2 = (1 - (1 - 2w_e/k)^{w_c - 1})/2$$

and

$$p_3 = \frac{1}{\binom{k}{w_u}\binom{k}{w_f}} \sum_{\substack{1 \le l \le \min(w_u, w_f) \\ l \text{ odd}}} \binom{k}{l}\binom{k-l}{w_u - l}\binom{k - w_u}{w_f - l}.$$

Proof. Write

$$(\mathbf{d}(x))_j = \left(\sum_{v \in \mathsf{supp}(\mathbf{c})} x^{-v}\mathbf{s}(x) \right)_j = \sum_{v \in \mathsf{supp}(\mathbf{c})} \left(x^{-v}\mathbf{s}(x)\right)_j$$

Fix the value of $v \in \mathsf{supp}(\mathbf{c})$ and consider the summand $(x^{-v}\mathbf{s}(x))_j$. We can expand it as in (2) as

$$(x^{-v}\mathbf{s}(x))_j = (\mathbf{e}(x))_j + (x^{-v}\mathbf{u}(x)\mathbf{f}(x))_j + \left(\sum_{l \in \mathsf{supp}(\mathbf{c}) \setminus \{v\}} x^{l-v}\mathbf{e}(x) \right)_j. \tag{8}$$

The random variable $(\mathbf{e}(x))_j$ is Bernoulli distributed with parameter

$$p_1 = w_e/k.$$

Note that $(x^{l-v}\mathbf{e})_j$ and $(\mathbf{e}(x))_j$ are identically distributed, as the coefficient vector of the former is the right circular shift of the coefficient vector of the latter, by $l - v$ positions. Therefore, from Lemma 2 we obtain that $\left(\sum_{l \in \mathsf{supp}(\mathbf{c}) \setminus \{v\}} x^{l-v}\mathbf{e}(x)\right)_j$ is Bernoulli distributed with parameter

$$p_2 = (1 - (1 - 2w_e/k)^{w_c - 1})/2.$$

Finally, from Lemma 1 and Remark 1 we have that the random variable $(x^{-v}\mathbf{u}(x)\mathbf{f}(x))_j$ is Bernoulli distributed with parameter

$$p_3 = \frac{1}{\binom{k}{w_u}\binom{k}{w_f}} \sum_{\substack{1\le l\le \min(w_u,w_f)\\ l \text{ odd}}} \binom{k}{l}\binom{k-l}{w_u-l}\binom{k-w_u}{w_f-l}.$$

Bearing in mind that we are considering events over $\mathbb{F}_2$, the sum of the three random variables is Bernoulli distributed with parameter

$$p = \prod_{i=1}^{3} p_i + \sum_{i=1}^{3} p_i \prod_{j\in[3], j\neq i} (1-p_j),$$

i.e. $(x^{-v}\mathbf{s}(x))_j$ succeeds if either only one or all the three variables produce success. □

At this point the proof of Theorem 1 is straightforward.

Proof of Theorem 1. The statement directly follows from Prop. 2 and the aforementioned assumptions of independence of the random variables involved in the sum. □

In the settings of our attack, we are going to specialize Theorem 1 to the case where $p_1 \in \{0,1\}$. This allows us to study the effect of the j-th coordinate of the vector $\mathbf{e}_i$ on the distribution of the j-th coordinate of $\mathbf{d}_i$. Notice that, in order to have a clear understanding of the behaviour of the j-th component of $\mathbf{d}(x)$, as defined in Eq. (3), we needed to sum the $(x^{-v}\mathbf{s}(x))_j$ random variables over all the $v \in \mathsf{supp}(\mathbf{c})$, where in this case the events are intended over $\mathbb{Z}$. While modelling this, we made the simplifying assumption that the random variables $x^{-v}\mathbf{u}_i(x)$ and $\sum_{l\in\mathsf{supp}(\mathbf{c})\setminus\{v\}} x^l\mathbf{e}(x)$ are independently distributed, as v varies in the support of $\mathbf{c}$. Experimental results in support of this assumption will be provided, showing that this assumption still manages to capture satisfactorily the behaviour of $\mathbf{d}(x)$, and, even more importantly, it does not affect the outcome of the cryptanalysis. Figure 4 gives a visual representation of how Theorem 1 approximates the behaviour of $\mathbf{d}_i$.

4.3 Recovering k Zero Bits of e

As mentioned in Sect. 4.1, we aim at finding k zero entries in $\mathbf{e}$. The columns of $\mathbf{H}$ corresponding to the remaining k positions will constitute a square $k \times k$ matrix which, in the case of being invertible, can be used to compute the entire value $\mathbf{e} = (\mathbf{e}_1, \mathbf{e}_2)$. The idea is to use the knowledge of the distribution of the entries $(\mathbf{d}_i)_j$, provided in Theorem 1, and find $\lceil k/2 \rceil$ error free positions in both the left side $\mathbf{e}_1$ and the right side $\mathbf{e}_2$ of $\mathbf{e}$, adding up to (at least) k error free positions. We search for an optimal threshold value τ such that the probability $\mathbb{P}((\mathbf{e}_i)_j = 0 \mid (\mathbf{d}_i)_j < \tau)$ is large enough. For our attack, we choose τ as follows:

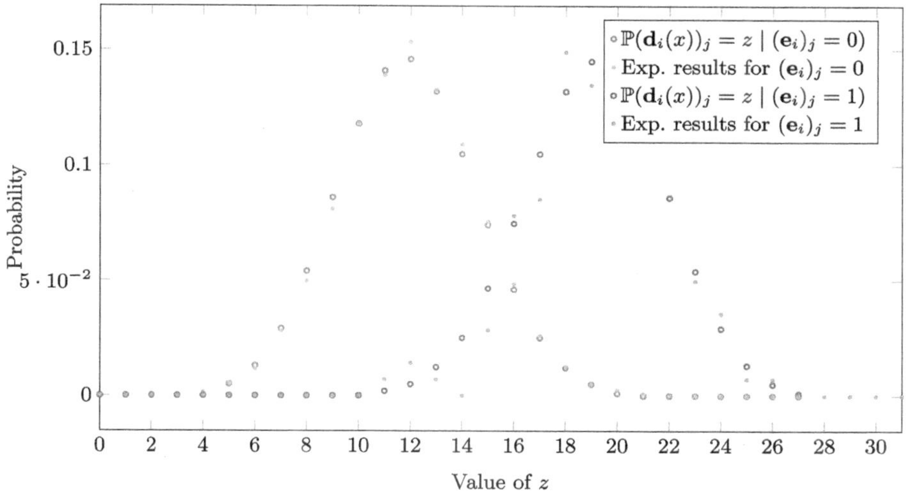

Fig. 4. In red (resp. dark gray) the theoretical estimates $\mathbb{P}(\mathbf{d}_i(x))_j = h \mid (\mathbf{e}_i)_j = 0)$ (resp. $\mathbb{P}(\mathbf{d}_i(x))_j = h \mid (\mathbf{e}_i)_j = 1)$), as h varies in the interval $[w_c]$, for security level 128. In orange (resp. light gray) the associated experimental results obtained for a random instance of HWQCS with 128 bit of security. (Color figure online)

$$\tau := \max\left\{h \in [w_c] \mid w_e \sum_{l=0}^{h} \mathbb{P}((\mathbf{d}_i)_j = l \mid (\mathbf{e}_i)_j = 1) < 1\right\}, \tag{9}$$

i.e. the largest integer such that, for each signature, the expected number of entries j such that $(\mathbf{e}_i)_j = 1$ and $(\mathbf{d}_i)_j < \tau$ is less than 1. This means that, on average, all the coordinates j such that $((\mathbf{d}_i)_j) < \tau$ will be error free. Let N_0 and N_1 be the expected number of error free positions and error positions that we can find by taking all j with $(\mathbf{d}_i)_j < \tau$, respectively. Then

$$N_0 := (k - w_e) \sum_{h=0}^{\tau-1} \mathbb{P}((\mathbf{d}_i)_j = h \mid (\mathbf{e}_i)_j = 0),$$

and

$$N_1 := (w_e) \sum_{h=0}^{\tau-1} \mathbb{P}((\mathbf{d}_i)_j = h \mid (\mathbf{e}_i)_j = 1).$$

Therefore, the probability that $(\mathbf{e}_i)_j = 0$ for all $j \in [k]$ such that $(\mathbf{d}_i)_j < \tau$ is given by

$$p_{succ} := \prod_{j \in [k], (\mathbf{d}_i)_j < \tau} \mathbb{P}((\mathbf{e}_i)_j = 0 \mid (\mathbf{d}_i)_j < \tau) = \left(\frac{N_0}{N_0 + N_1}\right)^{N_0 + N_1}.$$

The values for τ, N_0, N_1 and of p_{succ} for each suggested parameter set of HWQCS are given in Table 2. These values indicate how many error free bits of $\mathbf{e}_i$ we can expect to recover with our strategy. Surprisingly enough, for security level 256 we can reconstruct more than $\lceil k/2 \rceil$ error free positions of $\mathbf{e}_i$. The same does not hold for levels 128 and 192, meaning that we need to guess $\lceil k/2 \rceil - N_0$ error free positions on each side of $\mathbf{e}$.

Table 2. For each security level, we report the threshold value τ, the values N_0 and N_1 of zero and one bits that we expect to find among the j's such that $(\mathbf{d}_i)_j < \tau$, and the probability p_{succ} that this event occurs. The value $\lceil \frac{k}{2} \rceil - N_0$ is the number of positions that we still need to guess to fully reconstruct $\lceil k/2 \rceil$ entries.

Security level	$\lceil \frac{k}{2} \rceil$	τ	N_0	N_1	p_{succ}	$\lceil \frac{k}{2} \rceil - N_0$
128	6270	12	5564.3997	0.3995	0.6707	706
192	9459	15	7574.2963	0.2521	0.7771	1885
256	12709	21	13560.6279	0.3457	0.7077	-

Table 6 reports the probability distribution and expected values using our approximation of $(\mathbf{d}_i)_j$ given by Theorem 1.

Remark 2. We are considering $\lceil k/2 \rceil$ in order to make sure we are recovering enough positions as k is an odd number for all three parameter sets. This implies that the probabilities we are computing are actually underestimates. Also, for security 256 we might select a subset of $\lceil k/2 \rceil$ of the N_0 expected error free positions, which would increase the value of the success probability to $p_{succ} = 0.7233$. We will exploit this in Sect. 5.2 to give a speed-up of our attack.

Remark 3. The security level 256 would have values as reported in Table 3.

Table 3. Value of τ as given in Eq. (9), and the values of N_0 and N_1 of zero and one bits that we expect to find among the j's such that $(\mathbf{d}_i)_j < \tau$, as well as the probability p_{succ} that this event occurs.

Security level	$\lceil \frac{k}{2} \rceil$	τ	N_0	N_1	p_{succ}
256	12709	22	16336.7984	0.8193	0.4407

As mentioned at the beginning of this subsection, we are searching for a total of $\lceil k/2 \rceil$ error free entries in $\mathbf{e}_i$. The value of N_0 for security level 256 with $\tau = 22$, accounts for many more positions than actually needed, with somewhat smaller success probability compared to the levels 128 and 192. Decreasing the value of τ to 21 still gives N_0 larger than $\lceil k/2 \rceil$ and increases the success probability by more than 0.25.

We are left with the problem of finding the missing $\lceil \frac{k}{2} \rceil - N_0$ zero positions of $\mathbf{e}_i$ for security levels 128 and 192. According to the behaviour of the $(\mathbf{d}_i)_j$

entries, we aim at finding these values among the j's such that $(\mathbf{d}_i)_j = \tau$, i.e. the set of positions that have the least probability of containing an error, after those we already chose. Let M_0 and M_1 be the expected number of error free and error positions j such that $(\mathbf{d}_i)_j = \tau$, respectively. Then

$$M_0 = (k - w_e)\mathbb{P}((\mathbf{d}_i)_j = \tau \mid (\mathbf{e}_i)_j = 0)$$

and

$$M_1 = (w_e)\mathbb{P}((\mathbf{d}_i)_j = \tau \mid (\mathbf{e}_i)_j = 1).$$

Therefore the probability such that $\lceil \frac{k}{2} \rceil - N_0$ randomly chosen positions j such that $(\mathbf{d}_i)_j = \tau$ will be error free is given by

$$q_{succ} := \left(\frac{M_0}{M_0 + M_1} \right)^{\lceil \frac{k}{2} \rceil - N_0}.$$

The values of M_0, M_1 and q_{succ} for levels 128 and 192 are reported in Table 4.

Table 4. For security parameters 128 and 192, $\lceil \frac{k}{2} \rceil - N_0$ denotes the number of entries which is still necessary to guess to perform our attack. Values M_0 and M_1 represent the expected number of error free and error positions among the j's such that $(\mathbf{d}_i)_j = \tau$.

Security level	$\lceil \frac{k}{2} \rceil - N_0$	M_0	M_1	q_{succ}
128	706	1806.5588	0.7402	0.7491
192	1885	2433.8479	0.4371	0.7129

The overall probability of a correct recovery of $\lceil k/2 \rceil$ zero bits of $\mathbf{e}_i$ is then

$$p_{tot} := p_{succ} \cdot q_{succ}.$$

For the case of security level 256 we consider $q_{succ} = 1$ as $N_0 > \lceil \frac{k}{2} \rceil$. Applying this technique on both $\mathbf{e}_1$ and $\mathbf{e}_2$ we can recover k error free positions of $\mathbf{e}$ with probability of success p_{tot}^2. Values for each security level are reported in Table 5.

Table 5. For each security level, k and p_{tot}^2 are the number of error free positions we need to recover and the probability of recovering them correctly, respectively.

Security level	k	p_{tot}^2
128	12539	0.2524
192	18917	0.3070
256	25417	0.5008

In other words, Table 5 says that we can successfully recover k error free coordinates of $\mathbf{e}$ of around one fourth of the signatures of security level 128, slightly less than one third of those of level 192 and of half of the signatures of level 256. In the next section, we show how to completely reconstruct $\mathbf{e}$ using the k error free positions we recovered in this section.

Table 6. Values of the distribution of $(\mathbf{d}_i(x))_j$ for security level 128. For every h, we report the probability that $(\mathbf{d}_i(x))_j = h$, conditioning on the event $(\mathbf{e}_i)_j = 0, 1$. Two more columns report the expected number of entries j associated to $(\mathbf{d}_i(x))_j = h$.

h	$\mathbb{P}((\mathbf{d}_i(x))_j = h \mid (\mathbf{e}_i)_j = 0)$	Exp. number of entries	$\mathbb{P}((\mathbf{d}_i(x))_j = h \mid (\mathbf{e}_i)_j = 1)$	Exp. number of entries
0	$3.078 \cdot 10^{-7}$	0.004	$1.248 \cdot 10^{-13}$	$1.759 \cdot 10^{-11}$
1	$5.935 \cdot 10^{-6}$	0.074	$6.220 \cdot 10^{-12}$	$8.770 \cdot 10^{-10}$
2	$0.006 \cdot 10^{-2}$	0.686	$1.500 \cdot 10^{-10}$	$2.115 \cdot 10^{-8}$
3	$0.003 \cdot 10^{-1}$	4.128	$2.331 \cdot 10^{-9}$	$3.286 \cdot 10^{-7}$
4	0.001	17.973	$2.623 \cdot 10^{-8}$	$3.698 \cdot 10^{-6}$
5	0.005	60.371	$2.277 \cdot 10^{-7}$	$0.003 \cdot 10^{-2}$
6	0.013	162.726	$1.586 \cdot 10^{-6}$	$0.002 \cdot 10^{-1}$
7	0.029	361.500	$9.109 \cdot 10^{-6}$	0.001
8	0.054	674.586	$0.004 \cdot 10^{-2}$	0.006
9	0.086	1072.334	$0.001 \cdot 10^{-1}$	0.025
10	0.118	1467.441	$0.006 \cdot 10^{-1}$	0.090
11	0.141	1742.590	0.002	0.276
12	0.146	1806.558	0.005	0.740
13	0.132	1642.366	0.0123	1.739
14	0.105	1313.477	0.025	3.595
15	0.074	925.951	0.0464	6.550
16	0.046	575.964	0.0746	10.530
17	0.025	316.115	0.105	14.937
18	0.012	152.936	0.132	18.678
19	0.005	65.088	0.145	20.545
20	0.001	24.292	0.140	19.818
21	0.0006	7.914	0.118	16.688
22	0.0001	2.237	0.086	12.195
23	0.00004	0.544	0.054	7.671
24	$9.109 \cdot 10^{-6}$	0.112	0.029	4.11
25	$1.586 \cdot 10^{-6}$	0.019	0.013	1.850
26	$2.277 \cdot 10^{-7}$	0.002	0.005	0.686
27	$2.623 \cdot 10^{-8}$	$0.003 \cdot 10^{-1}$	0.001	0.204

4.4 Completing the Reconstruction of E

In this section, we exploit the analysis performed so far to fully reconstruct the ephemeral values $\mathbf{e}_1, \mathbf{e}_2$ with a certain probability. Recall that

$$\mathbf{b} = \mathbf{H}\mathbf{e}^\top, \tag{10}$$

where $\mathbf{e} = (\mathbf{e}_1, \mathbf{e}_2)$ and $\mathbf{H} = \big(\mathsf{circ}(\mathbf{h}), \mathsf{circ}(\mathbf{h})^{-1}\big)$.

Let $J \subset [2k]$ the set of k positions recovered using the strategy outlined in Sect. 4.3 and let $I := [2k] \backslash J$. Finally, let $\mathbf{H}_I$ be the submatrix of $\mathbf{H}$ which consists of the columns indexed by I. We can treat $\mathbf{H}_I$ as a random matrix in $\mathbb{F}_2^{k \times k}$. Assume that $\mathbf{H}_I$ is invertible. Then given the syndrome equation $\mathbf{b} = \mathbf{H}\mathbf{e}^\top$ of $\mathbf{e}$

as in Sign, we can compute $\bar{\mathbf{e}} = \mathbf{H}_I^{-1}\mathbf{b}$ and thus reconstruct $\mathbf{e} = (e_0, \ldots, e_{2k-1})$ as

$$e_h = \begin{cases} \bar{e}_h & \text{if} \quad h = i \text{ for some } i \in I \\ 0 & \text{otherwise,} \end{cases} \tag{11}$$

for each $h \in [2k]$. The probability for a random $k \times k$ matrix to be invertible is given by Proposition 2.

Proposition 2. *Let $\mathbf{H} \in \mathbb{F}_2^{k\times k}$ be a random square matrix. The probability that $\mathbf{H}$ is invertible is given by*

$$p_{inv} := \prod_{i=0}^{k-1}\left(1 - \frac{1}{2^{k-i}}\right). \tag{12}$$

Proof. For $k = 1$ the result is trivial. Suppose k is greater than one. If we have $k - 1$ independent vectors, the probability that the k-th vector is independent of the previous ones is $(2^k - 2^{k-1})/2^k$. Multiplying the probability for which these $k - 1$ vectors are independent and the probability that the k-th vector is independent of the others yields

$$\frac{2^k - 2^{k-1}}{2^k} \cdot \prod_{i=0}^{k-2}\left(1 - \frac{1}{2^{k-i}}\right) = \prod_{i=0}^{k-1}\left(1 - \frac{1}{2^{k-i}}\right).$$

For each set of suggested parameters of HWQCS, the value of the probability of $\mathbf{H}_I$ to be invertible is $p_{inv} := 0.2888$.

Remark 4. The code used to compute data for this section can be found at https://github.com/triki96/Cryptanalysis-of-HWQCS.

5 Universal Forgery

In this section, we exploit the reconstruction of $\mathbf{e}_1, \mathbf{e}_2$ to mount a universal forgery attack. Suppose an adversary $\mathcal{A}$ intercepts a signature $(\mathbf{c}, \mathbf{b}, \mathbf{s}_1, \mathbf{s}_2)$ used by a honest signer to sign a given message m. Note that:

$$\begin{aligned} \mathbf{b}(x) &= \mathbf{e}_1(x)\mathbf{h}(x) + \mathbf{e}_2(x)\mathbf{h}^{-1}(x), \\ \mathbf{c}(x) &= \mathsf{hash}_{w_c}(m, \mathbf{b}, \mathbf{u}_1(x)\mathbf{f}_2(x) - \mathbf{u}_2(x)\mathbf{f}_1(x), \mathsf{pk}), \\ \mathbf{s}_i(x) &= \mathbf{u}_i(x)\mathbf{f}_i(x) + \mathbf{c}(x)\mathbf{e}_i(x). \end{aligned}$$

Now, suppose the $\mathcal{A}$ is given a random message m', and it is asked to sign that message. The adversary needs $\sigma' = (\mathbf{c}', \mathbf{b}', \mathbf{s}_1', \mathbf{s}_2')$ which satisfies $\mathsf{Verify}(m', \mathsf{pk}, \sigma') = 1$. Therefore, $\mathcal{A}$ computes the signature in the following way:

$$\begin{aligned} \mathbf{b}'(x) &= \mathbf{b}(x), \\ \mathbf{c}'(x) &= \mathsf{hash}_{w_c}(m', \mathbf{b}', \mathbf{s}_1(x)\mathbf{h}(x) + \mathbf{s}_2(x)\mathbf{h}(x)^{-1} - \mathbf{c}(x)\mathbf{b}(x), \mathsf{pk}), \\ \mathbf{s}_i'(x) &= \mathbf{s}_i(x) - \mathbf{c}(x)\mathbf{e}_i(x) + \mathbf{c}'(x)\mathbf{e}_i(x) = \mathbf{u}_i(x)\mathbf{f}_i(x) + \mathbf{c}'(x)\mathbf{e}_i(x). \end{aligned}$$

The verifier computes

$$\begin{aligned}\mathbf{t}'(x) &= \mathbf{s}_1'(x)\mathbf{h}(x) + \mathbf{s}_2'(x)\mathbf{h}^{-1}(x) - \mathbf{c}'(x)\mathbf{b}'(x)\\ &= \mathbf{s}_1'(x)\mathbf{h}(x) + \mathbf{s}_2'(x)\mathbf{h}^{-1}(x) - \mathbf{c}'(x)\mathbf{b}(x)\\ &= (\mathbf{u}_1(x)\mathbf{f}_1(x) + \mathbf{c}'(x)\mathbf{e}_1(x))\mathbf{h}(x) + (\mathbf{u}_2(x)\mathbf{f}_2(x) + \mathbf{c}'(x)\mathbf{e}_2(x))\mathbf{h}^{-1}(x) +\\ &\quad - \mathbf{c}'(x)(\mathbf{e}_1(x)\mathbf{h}(x) + \mathbf{e}_2(x)\mathbf{h}^{-1}(x))\\ &= \mathbf{u}_1(x)\mathbf{f}_2(x) + \mathbf{u}_2(x)\mathbf{f}_1(x).\end{aligned}$$

Clearly $\mathbf{t}' \neq 0$ and $w(\mathbf{t}') \leq w_t$, as this is the same value computed in the verification phase of the first sign. It remains to check whether $\mathbf{c}'' := \mathsf{hash}_{w_c}(m', \mathbf{b}', \mathbf{t}', \mathsf{pk})$ is equal to $\mathbf{c}'$, but

$$\begin{aligned}\mathbf{c}'' &= \mathsf{hash}_{w_c}(m', \mathbf{b}', \mathbf{t}', \mathsf{pk})\\ &= \mathsf{hash}_{w_c}(m', \mathbf{b}, \mathbf{u}_1(x)\mathbf{f}_2(x) + \mathbf{u}_2(x)\mathbf{f}_1(x), \mathsf{pk})\\ &= \mathbf{c}'.\end{aligned}$$

This concludes the forgery.

Algorithm 5.1 is a probabilistic algorithm that sequentially analyses HWQCS signatures in order to retrieve the ephemeral values $\mathbf{e}_1, \mathbf{e}_2$ and forge a new signature for m'.

5.1 Complexity of Our Attack

We estimate the success probability of Algorithm 5.1. The attack succeeds at reconstructing the data needed for a universal forgery if we can correctly obtain k error free positions of $\mathbf{e}$ and if the matrix $\mathbf{H}_I$, which depends on the recovered positions, is invertible. We computed the success probabilities p_{tot}^2 and p_{inv} of both the events in Sects. 4.3 and 4.4, respectively. Therefore, the success probability of Algorithm 5.1 is given by the product

$$p_{break} := p_{tot}^2 \cdot p_{inv},$$

and thus, the expected number of attempts to achieve success is given by $\lceil 1/p_{break} \rceil$. The cost of each attempt is dominated by that of inverting the matrix $\mathbf{H}_I \in \mathbb{F}_2^{k \times k}$, which is in $\mathcal{O}(k^{2.37})$. Therefore, Algorithm 5.1 runs in time $\mathcal{O}(2^{\log(k^{2.37}/p_{break})})$. The values of p_{break}, the expected number of needed attempts and the cost of the attack for each security level of HWQCS are reported in Table 7.

We implemented an unoptimized version of our attack in SageMath and ran it on a Linux Mint virtual machine. The code stops after recomputing $\mathbf{e}'$ and checking it against the real error vector $\mathbf{e}$. If the two match, the forgery succeeds. Table 8 reports the average, on 10 runs of the attack for each security level, of the number of signatures needed to recover $\mathbf{e}$ and the average time consumed analysing each signature ().

Remark 5. The code for our attack can be found in Appendix A. More code concerning our analysis can be found at https://github.com/triki96/Cryptanalysis-of-HWQCS.

Algorithm 5.1. AttackHWQCS

Input $\mathsf{pk} = \mathbf{h} \in \mathbb{F}_2^k, \tau \in \mathbb{N}$ according to Table 2.
Output $\mathbf{e}_1, \mathbf{e}_2 \in V_{k,w_e}$ such that $\mathbf{b}(x) = \mathbf{e}_1(x)\mathbf{h}(x) + \mathbf{e}_2(x)\mathbf{h}^{-1}(x)$ for some valid signature pair (m, σ).

1. Compute $\mathbf{H} = \left(\mathsf{circ}(\mathbf{h}), \mathsf{circ}(\mathbf{h})^{-1}\right)$;
2. Request $(m, \sigma) = (m, (\mathbf{c}, \mathbf{b}, \mathbf{s}_1, \mathbf{s}_2))$;
3. Compute $\mathbf{d}_i(x) := \sum_{v \in \mathsf{supp}(\mathbf{c})} x^{-v}\mathbf{s}_i(x) \in \mathbb{Z}[x]$;
4. Set
   - $J_{1,1} := \{j \in \{1, \ldots, k\} \mid (\mathbf{d}_1(x))_j < \tau\}$ and
   - $J_{1,2} := \{j \in \{1, \ldots, k\} \mid (\mathbf{d}_2(x))_j < \tau\}$;
5. Set random
   - $J_{2,1} \subset \{j \in \{1, \ldots, k\} \mid (\mathbf{d}_1(x))_j = \tau\}$ of size $\lceil k/2 \rceil - \#J_{1,1}$ and
   - $J_{2,2} \subset \{j \in \{1, \ldots, k\} \mid (\mathbf{d}_2(x))_j = \tau\}$ of size $\lceil k/2 \rceil - \#J_{1,2}$
6. Set $I := \{1, \ldots, 2k\} \setminus (J_{1,1} \cup J_{1,2} \cup J_{2,1} \cup J_{2,2})$;
7. Set $\mathbf{H}_I$ the submatrix of $\mathbf{H}$ made by columns of $\mathbf{H}$ indexed by I;
8. If $\mathbf{H}_i$ is not invertible then go to (2);
9. Compute $\bar{\mathbf{e}} = \mathbf{H}_I^{-1}\mathbf{b}$;
10. Set $\mathbf{e}' = (e'_0, \ldots, e'_{2k-1}) = (\mathbf{e}'_1, \mathbf{e}'_2)$ as
$$e'_h = \begin{cases} \bar{e}_h & \text{if} \quad h = i \text{ for some } i \in I, \\ 0 & \text{otherwise.} \end{cases}$$
11. Let m' be a new message to be signed;
12. Compute $\mathbf{b}'(x) = \mathbf{b}(x)$;
13. Compute $\mathbf{c}'(x) = \mathsf{hash}_{w_c}(m', \mathbf{b}', (\mathbf{s}_1(x)\mathbf{h}(x) + \mathbf{s}_2(x)\mathbf{h}^{-1}(x) - \mathbf{c}(x)\mathbf{b}'(x), \mathbf{h})$;
14. Compute $\mathbf{s}'_i(x) = \mathbf{s}_i(x) - \mathbf{c}(x)\mathbf{e}'_i(x) + \mathbf{c}'(x)\mathbf{e}'_i(x)$;
15. If not $\mathsf{Verify}(m', \mathbf{h}, (\mathbf{c}', \mathbf{b}', \mathbf{s}'_1, \mathbf{s}'_2))$ then go to (2);
16. Return $\mathbf{e}_1, \mathbf{e}_2$ and σ.

5.2 BONUS: Breaking HWQCS Security Level 256 with One Signature

Due to the large value of N_0 as an effect of our choice of $\tau = 21$ for security level 256, see Table 2, it is possible to slightly improve on our attack by intercepting only one signature. Indeed, we can take advantage of this surplus of potentially error free positions that we obtain by taking all j such that $(\mathbf{d}_i(x))_j < \tau$. The idea is to randomly pick a subset J_i of cardinality $\lceil k/2 \rceil$ of the $N_0 + N_1$ positions such that $(\mathbf{d}_i(x))_j < \tau$ for both $i = 1, 2$. In other words, we are applying plain information set decoding on a potentially error free set of positions aiming at finding an invertible submatrix of $\mathbf{H}$.

Note that J_i has probability of being error free equal to $\left(\frac{N_0}{N_0+N_1}\right)^{\lceil k/2 \rceil}$. Since we are doing this for both left and right-hand side of $\mathbf{e}$, the overall probability of $J := J_1 \cup J_2$ of being error free is $p := \left(\frac{N_0}{N_0+N_1}\right)^k$. According to the values in Table 2, we have $p = 0.5231$. Set now $I_J := \{1, \ldots, 2k\} \setminus J$ and let $\mathbf{H}_{I_J}$ be

Table 7. For each security level, p_{break} denotes the probability that our attack successfully retrieves the ephemeral values $\mathbf{e}_1, \mathbf{e}_2$ from a given signature. The expected number of signatures for our attack to succeeds and the $\log_2$ of the cost of the attack are reported.

Security level	p_{break}	Number of signatures	Cost
128	0.0727	14	36.04
192	0.0887	12	37.25
256	0.1446	7	37.48

Table 8. Average number of signatures needed to mount our universal forgery attack, and average time needed to analyse each signature, on 10 runs of our attack for each security level.

Security level	Number of signatures	Consumed time (s)
128	17	40.33
192	9	92.50
256	3	133.25

the matrix consisting of the columns of $\mathbf{H}$ indexed by I_J. As per proposition 2, the probability of $\mathbf{H}_{I_J}$ of being invertible is $p_{\text{inv}} = 0.2888$. In the case that $\mathbf{H}_{I_J}$ is invertible, we continue the same way as in our attack, leading to a forgery. In the case that $\mathbf{H}_{I_J}$ is not invertible, we randomly pick other two subsets J_i of cardinality $\lceil k/2 \rceil$ of the $N_0 + N_1$ positions such that $(\mathbf{d}_i(x))_j < \tau$ for both $i = 1, 2$ and test $\mathbf{H}_{I_J}$.

The success probability of recovering the correct $\mathbf{e}$ becomes then

$$p_{\text{break}} = p \cdot p_{\text{inv}} = 0.1511,$$

which gives an improvement of 0.065 on the probability 0.1446 reported in Table 7. The expected number of attempts of picking random J_1 and J_2 and finding an invertible matrix $\mathbf{H}_I I_J$, with $J = J_1 \cup J_2$, is $\lceil 1/p_{\text{break}} \rceil = 5$. Observe that Algorithm 5.1 tests one matrix for each signature, and requests a new signature if the matrix is not full rank. This approach does not discard the signature, but discards J_1 and J_2 and picks two fresh sets. This is equivalent to tweaking Step 8 in Algorithm 5.1 to:

– 8. If $\mathbf{H}_I$ is not invertible, then go to (5).

In terms of number of matrix inversions this approach improves on the value reported in Table 7 by 2.

We can argue that, due to the gap $N_0 + N_1 - \lceil k/2 \rceil$, there exist at least 5 distinct choices of J_1 and J_2 columns such that each choice spans subspaces of dimensions J_1 and J_2, respectively, which are linearly independent, ensuring that H_{I_J} is full rank. This is corroborated by our software, which simulates this one-signature configuration at the 256 security level for HWQCS and successfully performs the forgery.

Remark 6. The SageMath code in Appendix A implements this *single signature* version of the attack.

6 Conclusion

Through a detailed analysis of the information leakage on ephemeral keys during the signing process, we demonstrated how partial secret data can be efficiently reconstructed from a limited number of signatures, regardless of the chosen security level (128, 192, or 256 bits). This enabled the development of a universal forgery attack that exploits this leakage to produce valid signatures for arbitrary messages. We proposed a probabilistic algorithm, implemented in SageMath, which empirically supports the effectiveness and efficiency of our attack strategy, particularly at the 256-bit security level, where an additional optimization further accelerates the recovery of secret data.

Appendix A

```
from sage.doctest.util import Timer

# Level 128
#lam,k,wf,wu,we,wc,ws,wt,tau = 128,12539,145,33,141,31,4844, 4937,12

# Level 192
#lam,k,wf,wu,we,wc,ws,wt,tau = 192,18917,185,41,177,39,7450,7592,15

# Level 256
lam,k,wf,wu,we,wc,ws,wt,tau = 256,25417,145,51,191,51,10111,10216,21


F = GF(2)
R1.<y> = F[]
R.<x> = R1.quo(y^k - 1)

def weight(v):
return sum(1 for i in v if i==1)

def getmultmatrix(R,P2,nrows):
M = []
for i in range(k):
el = list((R*x^i).mod(P2))
M.append(el + [0 for j in range(k-len(el))])
M = matrix(Fqm,M)
```

```
return M[:nrows,:]

def keygen():
f = []
for i in range(2):
inv = False
while not inv:
P = Permutations(k)
p = P.random_element()
meta_pol = [1 for i in range(wf)] + [0 for i in range(k-wf)]
meta_pol = [meta_pol[p[i]-1] for i in range(k)]
fi = R(meta_pol)
inv = fi.is_unit()
f += [fi]

return f[0].inverse_of_unit()*f[1], f


def sign(H,sk):
while True:
e = []
u = []
P = Permutations(k)
for i in range(2):
p = P.random_element()
ei = [1 for i in range(we)] + [0 for i in range(k-we)]
ei = [ei[p[i]-1] for i in range(k)]
e += [ei]
for i in range(2):
p = P.random_element()
ui = [1 for i in range(wu)] + [0 for i in range(k-wu)]
ui = [ui[p[i]-1] for i in range(k)]
u += [ui]
e = vector(e[0]+e[1])
b = H*e

p = P.random_element()
c = [1 for i in range(wc)] + [0 for i in range(k-wc)]
c = [c[p[i]-1] for i in range(k)]
s = [R(u[i])*sk[i]+ R(c)*R(list(e)[k*i:k*(i+1)]) for i in range(2)]
if weight(s[1]) <= ws and weight(s[0]) <= ws and
            weight(R(u[0])*sk[1] + R(u[1])*sk[0]) <= wt:
return s,b,c,e

pk,sk = keygen()
```

```
print("computing public matrix, might take a while ...")
h1 = matrix.circulant(list(pk))
H = h1.augment(h1.inverse())
print("done")

times = []
numsig = []
timer = Timer()
for i in range(1):
broken = False
count = 1;
time = 0
timer.start()
while not broken:
print(f"generate one signature and analyze it ... {count}")
count +=1
s,b,c,e = sign(H,sk)
c = R1(c)
stats0 = [ZZ(0)]*k
stats1 = [ZZ(0)]*k
p0s0=[]
p1s0=[]
p0s1=[]
p1s1=[]
for m in c.monomials():
v = m.degree()
s0v = s[0]*x^(k-v)
s1v = s[1]*x^(k-v)
ls0 = list(s0v)
ls1 = list(s1v)
for i in range(k):
stats0[i] += ZZ(ls0[i])
stats1[i] += ZZ(ls1[i])

s0zeros = []
s0ones = []
s1zeros = []
s1ones = []
for i in range(k):
if e[i] == 0:
p0s0 += [i]
s0zeros += [stats0[i]]

else:
p1s0 += [i]
```

```
s0ones += [stats0[i]]

if e[k+i] == 0:
p0s1 += [i]
s1zeros += [stats1[i]]

else:
p1s1 += [i]
s1ones += [stats1[i]]



z0 = [i for i in range(k) if stats0[i]<tau]
print(len(z0),min(s0ones),len([i for i in s0ones if i<=tau]))

z1 = [k+i for i in range(k) if stats1[i]<tau]
print(len(z1),min(s1ones),len([i for i in s1ones if i<=tau]))

z013 = [i for i in range(k) if stats0[i]==tau]

z113 = [k + i for i in range(k) if stats1[i]==tau]

##### SEARCH  INVERTIBLE MATRIX
print("Start searching for an invertible submatrix")
Htmp = zero_matrix(k)
cols = []
if lam != 256:
print("guess some/other columns to exclude on the left and on the right..
s0smpcoord = sample(range(len(z013)), ceil(k/2)-len(z0))
s0smp = [z013[i] for i in s0smpcoord]
cols0 = z0 + s0smp
s1smpcoord = sample(range(len(z113)), k-ceil(k/2)-len(z1))
s1smp = [z113[i] for i in s1smpcoord]
cols = cols0 + z1 + s1smp
print(len(z0), len(s0smp), len(z1),len(s1smp))
sel = [i for i in range(2*k) if i not in cols]
print("build the submatrix ...")
Htmp = H[:,sel]
print(f'Matrix is invertible : {Htmp.rank()==k}')
if Htmp.rank() != k : continue
print("Computing error vector...")
etmp = Htmp.inverse()*b
erec = [0]*(2*k)
for i in range(k):
erec[sel[i]] = etmp[i]
```

```
broken = vector(erec) == e
print(f'Found error vector is valid : {broken}')
if broken:
timer.stop()
times += [timer.walltime]
numsig += [count]

else:
while not broken:
cols = []
coord = sample(range(len(z0+z1)), k)
for i in coord:
cols += [(z0 + z1)[i]]
print("256 -- ",len(cols))
sel = [i for i in range(2*k) if i not in cols]
print("build the submatrix ...")
Htmp = H[:,sel]
print(f'Matrix is invertible : {Htmp.rank()==k}')
if Htmp.rank() != k : continue
print("Computing error vector...")
etmp = Htmp.inverse()*b
erec = [0]*(2*k)
for i in range(k):
erec[sel[i]] = etmp[i]
broken = vector(erec) == e
print(f'Found error vector is valid : {broken}')
if broken:
timer.stop()
times += [timer.walltime]
numsig += [count]
```

References

1. Aragon, N., Baldi, M., Deneuville, J.-C., Khathuria, K., Persichetti, E., Santini, P.: Cryptanalysis of a code-based full-time signature. Des. Codes Crypt. **89**(9), 2097–2112 (2021). https://doi.org/10.1007/s10623-021-00902-7
2. Augot, D., Finiasz, M., Sendrier, N.: A family of fast syndrome based cryptographic hash functions. In: Dawson, E., Vaudenay, S. (eds.) Mycrypt 2005. LNCS, vol. 3715, pp. 64–83. Springer, Heidelberg (2005). https://doi.org/10.1007/11554868_6
3. Baldi, M., Bianchi, M., Chiaraluce, F., Rosenthal, J., Schipani, D.: Using LDGM codes and sparse syndromes to achieve digital signatures. In: Gaborit, P. (ed.) PQCrypto 2013. LNCS, vol. 7932, pp. 1–15. Springer, Heidelberg (2013). https://doi.org/10.1007/978-3-642-38616-9_1
4. Baldi, M., Deneuville, J.-C., Persichetti, E., Santini, P.: Cryptanalysis of a code-based signature scheme based on the Schnorr-Lyubashevsky framework. IEEE Commun. Lett. **25**(9), 2829–2833 (2021)

5. Courtois, N.T., Finiasz, M., Sendrier, N.: How to achieve a McEliece-based digital signature scheme. In: Boyd, C. (ed.) ASIACRYPT 2001. LNCS, vol. 2248, pp. 157–174. Springer, Heidelberg (2001). https://doi.org/10.1007/3-540-45682-1_10
6. Dallot, L.: Towards a concrete security proof of Courtois, Finiasz and Sendrier signature scheme. In: Lucks, S., Sadeghi, A.-R., Wolf, C. (eds.) WEWoRC 2007. LNCS, vol. 4945, pp. 65–77. Springer, Heidelberg (2008). https://doi.org/10.1007/978-3-540-88353-1_6
7. Damgård, I.B.: A design principle for Hash functions. In: Brassard, G. (ed.) CRYPTO 1989. LNCS, vol. 435, pp. 416–427. Springer, New York (1990). https://doi.org/10.1007/0-387-34805-0_39
8. Debris-Alazard, T., Sendrier, N., Tillich, J.-P.: Wave: a new family of trapdoor one-way preimage sampleable functions based on codes. In: Galbraith, S.D., Moriai, S. (eds.) ASIACRYPT 2019. LNCS, vol. 11921, pp. 21–51. Springer, Cham (2019). https://doi.org/10.1007/978-3-030-34578-5_2
9. Debris-Alazard, T., Tillich, J.-P.: Two attacks on rank metric code-based schemes: RankSign and an IBE scheme. In: Peyrin, T., Galbraith, S. (eds.) ASIACRYPT 2018. LNCS, vol. 11272, pp. 62–92. Springer, Cham (2018). https://doi.org/10.1007/978-3-030-03326-2_3
10. Deneuville, J.-C., Gaborit, P.: Cryptanalysis of a code-based one-time signature. Des. Codes Crypt. **88**(9), 1857–1866 (2020). https://doi.org/10.1007/s10623-020-00737-8
11. D'Alconzo, G., Meneghetti, A., Piasenti, P.: Security issues of CFS-like digital signature algorithms. J. Discrete Math. Sci. Crypt. **27**(1), 175–187 (2024)
12. Fiat, A., Shamir, A.: How To prove yourself: practical solutions to identification and signature problems. In: Odlyzko, A.M. (ed.) CRYPTO 1986. LNCS, vol. 263, pp. 186–194. Springer, Heidelberg (1987). https://doi.org/10.1007/3-540-47721-7_12
13. Fukushima, K., Roy, P.S., Xu, R., et al.: Random code-based signature scheme (RaCoSS). In: First Round Submission to the NIST Post-Quantum Cryptography Call (2017)
14. Gaborit, P., Ruatta, O., Schrek, J., Zémor, G.: RankSign: an efficient signature algorithm based on the rank metric. In: Mosca, M. (ed.) PQCrypto 2014. LNCS, vol. 8772, pp. 88–107. Springer, Cham (2014). https://doi.org/10.1007/978-3-319-11659-4_6
15. Hörmann, F., van Woerden, W.: FuLeakage: breaking FuLeeca by learning attacks. In: Reyzin, L., Stebila, D. (eds.) Advances in Cryptology – CRYPTO 2024. CRYPTO 2024. LNCS, vol. 14925. Springer, Cham (2024). https://doi.org/10.1007/978-3-031-68391-6_8
16. Kabatianskii, G., Krouk, E., Smeets, B.: A digital signature scheme based on random error-correcting codes. In: Darnell, M. (ed.) Cryptography and Coding 1997. LNCS, vol. 1355, pp. 161–167. Springer, Heidelberg (1997). https://doi.org/10.1007/BFb0024461
17. Kim, J.L., Hong, J., Lau, T.S.C., Lim, Y., Won, B.S.: REDOG and its performance analysis. Cryptology ePrint Archive (2022)
18. Li, Z., Xing, C., Yeo, S.L.: A new code based signature scheme without trapdoors. Cryptology ePrint Archive (2020)
19. Lyubashevsky, V.: Fiat-Shamir with aborts: applications to lattice and factoring-based signatures. In: Matsui, M. (ed.) ASIACRYPT 2009. LNCS, vol. 5912, pp. 598–616. Springer, Heidelberg (2009). https://doi.org/10.1007/978-3-642-10366-7_35

20. Melchor, C.A., et al.: Hamming quasi-cyclic (HQC). NIST PQC Round **2**, 4–13 (2018)
21. Merkle, R.C.: One way Hash functions and DES. In: Brassard, G. (ed.) CRYPTO 1989. LNCS, vol. 435, pp. 428–446. Springer, New York (1990). https://doi.org/10.1007/0-387-34805-0_40
22. Otmani, A., Tillich, J.-P.: An efficient attack on all concrete KKS proposals. In: Yang, B.-Y. (ed.) PQCrypto 2011. LNCS, vol. 7071, pp. 98–116. Springer, Heidelberg (2011). https://doi.org/10.1007/978-3-642-25405-5_7
23. Persichetti, E.: Efficient one-time signatures from quasi-cyclic codes: a full treatment. Cryptography **2**(4), 30 (2018)
24. Phesso, A., Tillich, J.-P.: An efficient attack on a code-based signature scheme. In: Takagi, T. (ed.) PQCrypto 2016. LNCS, vol. 9606, pp. 86–103. Springer, Cham (2016). https://doi.org/10.1007/978-3-319-29360-8_7
25. Picozzi, C., Meneghetti, A., Tognolini, G.: A post-quantum digital signature scheme from QC-LDPC codes. Cryptology ePrint Archive (2022)
26. Prabowo, T.F., Tan, C.H.: Attack on a code-based signature scheme from QC-LDPC codes. In: Codes, Cryptology and Information Security: 4th International Conference, C2SI 2023, Rabat, Morocco, May 29–31, 2023, Proceedings, pp. 136–149. Springer (2023)
27. Ren, F., Zheng, D., Wang, W., et al.: An efficient code based digital signature algorithm. Int. J. Netw. Secur. **19**(6), 1072–1079 (2017)
28. Ritterhoff, S., et al.: FuLeeca: a Lee-based signature scheme. Cryptology ePrint Archive (2023)
29. Santini, P., Baldi, M., Chiaraluce F.: Cryptanalysis of a one-time code-based digital signature scheme. In: 2019 IEEE International Symposium on Information Theory (ISIT), pp. 2594–2598. IEEE (2019)
30. Song, Y., Huang, X., Mu, Y., Wu, W., Wang, H.: A code-based signature scheme from the Lyubashevsky framework. Theoret. Comput. Sci. **835**, 15–30 (2020)
31. Tan, C.H., Prabowo, T.F.: High weight code-based signature scheme from QC-LDPC CODES. In: Seo, H., Kim, S. (eds.) Information Security and Cryptology - ICISC 2023. LNCS, vol. 14561, pp. 306–323. Springer, Singapore (2024)
32. Weger, V., Gassner, N., Rosenthal, J.: A survey on code-based cryptography. arXiv preprint arXiv:2201.07119 (2022)
33. Xagawa, K.: Practical attack on RaCoSS-r. Cryptology ePrint Archive (2018)

Increasing Index Sizes for Information Set Decoding Algorithms

Zachary Welch(✉)

Carleton University, Ottawa, ON K1S 5B6, Canada
zacharywelch@cmail.carleton.ca

Abstract. The security of code-based cryptography is largely based on the effectiveness of information set decoding (ISD) algorithms. This paper introduces a modification to Stern's algorithm that reduces average decryption time by at least 14.2% for a code of length 1024, dimension 524 capable of correcting 50 errors. We accomplish this by adding 3 extra columns to the columns reduced to the identity matrix and performing extra row reductions that manipulate the matrix into a form that still allows Stern's algorithm to be run.

Keywords: Code-Based Cryptography · Information Set Decoding · Stern's Algorithm · Row Reduction

1 Introduction

Code-based cryptosystems are cryptosystems based on the difficulty of decoding a random linear code. These cryptosystems often work by creating a linear code with some structure that allows for fast decoding and obfuscating that structure to be indistinguishable from a random linear code.

The first code-based cryptosystem was proposed by McEliece in 1978 [1] after it was shown that decoding a random linear code is NP-complete [2]. However, due to the large public-key size of this cryptosystem, it did not see widespread use compared to other public-key cryptosystems, often based on the factoring or discrete logarithm problems. While there are no known algorithms to quickly solve these problems on a classical computer, in 1994 it was shown that both factoring and discrete logarithm are vulnerable to an attack by a quantum adversary [3].

In response to recent advancements in quantum computing, standardization of post-quantum cryptosystems has been ongoing by several groups, most notably by NIST [4] and PQCrypto [5]. Code-based cryptosystems are a common type of cryptosystem in these competitions, including three of the four candidates in the 4th round of the NIST competition: Classic McEliece [6], BIKE [7] and HQC [8]. The fourth round 4 candidate SIKE [9] was broken [10] and is considered insecure, so all remaining candidates in round 4 are code-based cryptosystems.

V. Weger et al. (Eds.): CBCrypto 2024, LNCS 15531, pp. 94–109, 2025.
https://doi.org/10.1007/978-3-031-90229-1_5

The fastest known attacks against these cryptosystems are based on information set decoding (ISD). Information set decoding was originally proposed as a decoding algorithm over 60 years ago by Prange [11] and has seen numerous improvements since then [12–23].

In this paper we give a new algorithm that is capable of reducing average decryption time of Stern's algorithm by reducing the average number of iterations these algorithms need to perform to find a solution. The effect of this algorithm varies by both system and attack parameters, but we give one example which achieves at minimum a 14.2% reduction in decryption time for a code of length 1024, dimension 524 that is capable of correcting 50 errors. This algorithm does not lead to an asymptotic improvement and is less effective against systems with larger parameters.

2 Preliminaries

2.1 Notation

We let $\mathbb{F}_2$ denote the finite field of size 2.

We use lowercase bold letters to represent row vectors; when we need to discuss column vectors, we do so using the transpose operation, so $\boldsymbol{a}^T$ would be a column vector.

We use $\boldsymbol{a}_i$ to refer to the ith component of the vector $\boldsymbol{a}$ and $A_{i,j}$ to refer to the component in i^{th} row and j^{th} column of the matrix A. We let A_c denote the cth column of the matrix A and $A[c]$ denote the cth row of the matrix A.

2.2 Hamming Weight Error Correcting Codes

Definition 1. *A **linear code** is a k-dimensional subspace of $\mathbb{F}^n$ for some field $\mathbb{F}$ and some integers k and n.*

Definition 2. *A **parity-check matrix** of a linear code C is a $(n-k)\times n$ matrix over $\mathbb{F}$ of full rank such that $C = \{\boldsymbol{x} : H\boldsymbol{x}^T = \boldsymbol{0}^T\}$.*

For a given linear code the parity-check matrix is not unique. Let H be a parity-check matrix for some linear code C, then, for any invertible $(n-k)\times(n-k)$ matrix T, TH is also a parity-check matrix for C. Notably this means that performing row reduction operations on a parity-check matrix always produces a parity-check matrix for the same code.

Definition 3. *The **Hamming weight** of a vector $\boldsymbol{x} \in \mathbb{F}^n$, denoted $|\boldsymbol{x}|$, is the number of non-zero components of $\boldsymbol{x}$.*

Definition 4. *The **Hamming distance** between two vectors $\boldsymbol{x}, \boldsymbol{y} \in \mathbb{F}^n$ is $|\boldsymbol{x}-\boldsymbol{y}|$.*

Definition 5. *A linear code C is called a **Hamming weight error correcting code** capable of correcting t errors if, given $\boldsymbol{x}$ such that $\boldsymbol{x} = \mathbf{c} + \mathbf{e}$ with $\mathbf{c} \in C$ and $|\boldsymbol{e}| \leq t$, there exists an algorithm to find $\mathbf{c}$.*

Note that $\mathbf{c}$ must be the only vector in C within a Hamming distance of t to $\boldsymbol{x}$ to find it. The classical decoding problem for a linear code is, given $\boldsymbol{x} = \mathbf{c} + \boldsymbol{e}$, to find $\mathbf{c}$. This is equivalent to, for a given parity-check matrix H and vector $\boldsymbol{s}^T = H\boldsymbol{e}^T$ such that $|\boldsymbol{e}| \leq t$, find $\boldsymbol{e}$.

3 Stern's Algorithm

We start with a description of the ISD attack described by Stern [12] which forms the basis for modern ISD; Algorithm 1 is the version introduced in [12].

Let $H \in \mathbb{F}_2^{(n-k)\times n}$ be a parity-check matrix of a code of length n and dimension k that can correct t errors. Suppose we are given H and $H\boldsymbol{e}^T = \boldsymbol{s}^T$ with $\boldsymbol{e}$ having Hamming weight at most t and our task is to find $\boldsymbol{e}$, or equivalently a set of at most t columns of H that sum to $\boldsymbol{s}$. Stern's algorithm functions as follows:

Randomly select a set Z of $n-k$ linearly independent columns of H and replace H with the row reduction of H that gives the identity submatrix in these columns, performing the same row reductions on $\boldsymbol{s}$ to maintain the $H\boldsymbol{e}^T = \boldsymbol{s}^T$ equation. Split the columns of H not in Z into two subsets X and Y, with each column included in either X or Y independently with probability $\frac{1}{2}$.

For attack parameters $\ell \leq n-k$ and $p \leq \frac{t}{2}$, select a set $J \subset \{1,\ldots,n-k\}$. For every subset $A \subset X$ with p elements, compute the sum of columns in A over rows of H indexed by J and for every subset of $B \subset Y$ with p elements, compute the sum of $\boldsymbol{s}^T$ and columns in B over the rows indexed by J. For every pair of sets A and B with identical sums over the rows indexed by J, compute $\boldsymbol{w}^T$, the sum of $\boldsymbol{s}^T$ and all columns in $A \cup B$ over all rows of H. If the Hamming weight of $\boldsymbol{w}$ is less than or equal to $t-2p$, then the columns of Z with pivots in the same row as the 1's in $\boldsymbol{w}$ along with the columns of $A \cup B$ are the columns that sum to $\boldsymbol{s}$. If no such subsets A and B exist, then randomly select a new set Z and try again.

Algorithm 1 *Stern's Algorithm*

Input: Positive integers t, ℓ and p, matrix $H \in \mathbb{F}_2^{(n-k)\times n}$ and vector $\boldsymbol{s}$ such that $\boldsymbol{s}^T = H\boldsymbol{e}^T$ for some vector $\boldsymbol{e}$ with Hamming weight at most t.
Output: A set of at most t columns of H that sum to $\boldsymbol{s}^T$.

1 Set Z to a randomly selected set of $n-k$ linearly independent columns of H.
2 Set H to the row reduction of H that has the identity submatrix in the columns of Z.
3 Perform the same row reductions from step 2 on $\boldsymbol{s}^T$.
4 Set J to a random size ℓ subset of $\{1,\ldots,n-k\}$.
5 Set $H' \in \mathbb{F}_2^{|J|\times n}$ to the submatrix of H consisting of the rows indexed by J.
6 Set $\bar{\boldsymbol{s}}^T$ to the entries of $\boldsymbol{s}^T$ corresponding to rows indexed by J.
7 Split the columns of H not in Z into sets X and Y. Each column is added to X or Y with 50% probability.
8 For all size p subsets $A \subset X$
9   Index each set with $\boldsymbol{x}_A = \sum_{d\in A} H'_d$
10 For all size p subsets $B \subset Y$:
11   Compute $\boldsymbol{y}_B = \sum_{d\in B} H'_d + \bar{\boldsymbol{s}}^T$
12   For all indexed subsets A such that $\boldsymbol{x}_A = \boldsymbol{y}_B$:
13 $\boldsymbol{w}^T := \boldsymbol{s}^T + \sum_{d\in A\cup B} H_d$
14   If $|\boldsymbol{w}| \leq t-2p$:
15   Set C as the subset of Z with $1's$ in the same rows as $\boldsymbol{w}^T$.
16   Return $A \cup B \cup C$.
17 Go to step 1.

To see the idea behind Stern's algorithm, suppose H is written in the form

$$\begin{bmatrix} I_{n-k-\ell} & H_1 \\ 0 & H_2 \end{bmatrix}$$

for some binary matrices H_1 and H_2. Next split the vector $\boldsymbol{e} = (\dot{\boldsymbol{e}}, \bar{\boldsymbol{e}})$, where $\dot{\boldsymbol{e}}$ denotes the first $n-k-\ell$ components of $\boldsymbol{e}$ and $\bar{\boldsymbol{e}}$ denotes the last $k+\ell$ components of $\boldsymbol{e}$. Similarly decompose the vector $\boldsymbol{s}$ into $(\dot{\boldsymbol{s}}, \bar{\boldsymbol{s}})$, where $\dot{\boldsymbol{s}}$ denotes the first $n-k-\ell$ components of $\boldsymbol{s}$ and $\bar{\boldsymbol{s}}$ denotes the last ℓ components of $\boldsymbol{s}$.

We can decompose $H\boldsymbol{e}^T$ to find that $\dot{\boldsymbol{e}}^T + H_1\bar{\boldsymbol{e}}^T = \dot{\boldsymbol{s}}^T$ and $H_2\bar{\boldsymbol{e}} = \bar{\boldsymbol{s}}$; this first equation is equivalent to $\dot{\boldsymbol{s}}^T + H_1\bar{\boldsymbol{e}} = \dot{\boldsymbol{e}}$. Stern's algorithm attempts to find low Hamming weight $\boldsymbol{e}''$ such that $H_2\bar{\boldsymbol{e}} = \bar{\boldsymbol{s}}$ and $\dot{\boldsymbol{s}}^T + H_1\bar{\boldsymbol{e}}$ has low Hamming weight.

The Hamming weight of $\boldsymbol{e}$ is equal to the sum of the Hamming weights of $\boldsymbol{e}'$ and $\boldsymbol{e}''$; so a low Hamming weight $\boldsymbol{e}'$ and $\boldsymbol{e}''$ correspond to a low Hamming weight $\boldsymbol{e}$. The parameter p defines how the Hamming weight is distributed between $\boldsymbol{e}'$ and $\boldsymbol{e}''$.

3.1 Stern's Algorithm Logic

Let the subset of columns of Z that have a 1 in a rows indexed by J be called the columns corresponding to J and denote this set J_{col}. As a reminder, we are in $\mathbb{F}_2$, so addition and subtraction are the same operation, and $a+b=c$ implies $a=b+c$.

Theorem 2. [12, Proposition 2] *Suppose for some size t subset E of the columns of H, we have that $\sum_{d\in E} H_d = \boldsymbol{s}^T$. An iteration of Stern's algorithm identifies the set E if and only if there are exactly p columns in $X\cap E$, p columns in $Y\cap E$ and no columns in $J_{col}\cap E$.*

Theorem 3. [12, Theorem 1] *Suppose there exists a single size t subset E of the columns of H such that $\sum_{d\in E} H_d = \boldsymbol{s}^T$. The probability that a random selection of Z, X and Y finds E is:*

$$\frac{\binom{t}{2p}\binom{n-t}{k-2p}\binom{2p}{p}\binom{n-k-t+2p}{|J|}}{\binom{n}{k}4^p\binom{n-k}{|J|}}.$$

These theorems are proven in [12] and we skip the proofs here.

From the above two theorems, it seems that we would want to take $|J|=\ell$ as small as possible. While minimizing ℓ would reduce the number of iterations needed for Stern's algorithm to find an error vector, a smaller ℓ gives more subsets A and B such that $\sum_{a\in A} H'_a = \bar{\boldsymbol{s}}^T + \sum_{b\in B} H'_b$ in each iteration. Increasing ℓ by 1 halves the average number of subsets that need to be checked; optimal choice of ℓ requires the extra time checking subsets to be balanced against the reduced chance for an iteration of Stern's algorithm succeeding.

Increasing the parameter p gives more subsets of X and Y that need to be checked, so p is often left much smaller than the sizes of Z, X and Y would suggest. To give an example of this, [15, page 13] uses $p = 2$ and $\ell = 20$ for $n = 1024, k = 524$ and $t = 50$.

4 Increasing Index Sizes

It is possible to modify Stern's algorithm to increase the size of Z while still being able to identify if $\boldsymbol{w}^T = \boldsymbol{s}^T + \sum_{d \in A \cup B} H_d$ corresponds to a possible solution. As part of Stern's algorithm, the only property we used of Z when identifying potential subsets A and B was that the sum of $t-2p$ columns of Z had Hamming weight $t - 2p$. If, for some positive integer u, all columns of Z had Hamming weight at most u then the sum of $t - 2p$ columns of Z would have Hamming weight at most $u(t - 2p)$. It is possible to use this $u(t - 2p)$ in place of $t - 2p$ when checking the Hamming weight of $\boldsymbol{w}^T$.

When the Hamming weight of columns of Z is restricted to 1, the number of columns in Z is restricted to the number of available rows. If the Hamming weight of these columns is increased, the restriction of the number of columns in Z also increases, reducing the average number of iterations Stern's algorithm needs on average to find a solution.

The formula given in Theorem 3 assumes that $|Z| = n-k$ and the remaining k columns are split randomly between X and Y. By increasing $|Z|$ by some integer m, we can calculate the average number of iterations as the inverse of the formula given in 3 with k replaced by $k - m$. Below we give the average number of iterations for Stern's algorithm to find a solution if $|Z|$ is increased by 1, 3, and 10 using $\ell = 20$ and $p = 2$ on various parameter sets (Table 1).

Table 1. The average number of iterations Stern's algorithm needs to find a solution if the size of Z is increased.

Parameter Set	Default Size	+1 Column	+3 Columns	+10 Columns
Initial McEliece [1]	$7.350 \cdot 10^{11}$	$6.735 \cdot 10^{11}$	$5.612 \cdot 10^{11}$	$3.094 \cdot 10^{11}$
McEliece348864 [6]	$9.541 \cdot 10^{35}$	$8.79 \cdot 10^{35}$	$7.462 \cdot 10^{35}$	$4.223 \cdot 10^{35}$
McEliece6688128 [6]	$9.907 \cdot 10^{70}$	$9.167 \cdot 10^{70}$	$7.851 \cdot 10^{70}$	$4.571 \cdot 10^{70}$
McEliece8192128 [6]	$8.221 \cdot 10^{81}$	$7.606 \cdot 10^{81}$	$6.512 \cdot 10^{81}$	$3.786 \cdot 10^{81}$

While the number of iterations decreases, increasing the size of Z has the effect of making each iteration take more time. These effects are described in Sect. 5, and a detailed analysis of their effects is given in Sect. 7.

4.1 Row Reducing Three Additional Columns

For this modification to Stern's algorithm, we need an algorithm for row reducing H such that columns of Z each have weight at most u.

For the sake of notation, we assume that Z is the first $n - k + 3$ columns of H.

Algorithm 4 *Row reducing 3 additional columns*

Input: A binary $(n-k) \times n$ matrix H in RREF with its first $n-k$ columns linearly independent.
Output: A binary $(n-k) \times n$ matrix H' with the Hamming weight of the first $n-k+3$ columns of H at most 2 for all but 6 columns, those 6 columns have a Hamming weight of at most 3.

1 for all $i \in \{0,1\}^3$:
2   Set $r_i = -1$
3 for d from 1 to $n-k$:
4   Let $i =$concat$(H_{d,n-k+1}, H_{d,n-k+2}, H_{d,n-k+3})$
5   if $i = 000$:
6         $H'[d] = H[d]$
7   if $i \neq 000$:
8   if $r_i \neq -1$:
9             $H'[r_i] := H[r_i] + H[d]$
10         $r_i := d$
11 for all $i \in \{0,1\}^3$:
12     $H'[r_i] := H[r_i]$
13 $H'[r_{111}] := H[r_{111}] + H[r_{110}]$
14 $H'[r_{101}] := H[r_{101}] + H[r_{001}]$
15 $H'[r_{011}] := H[r_{011}] + H[r_{010}]$
16 return H'

Starting with a matrix in RREF, Algorithm 4 iterates through each row of H; cancelling out each row with more than a single 1 with the next row that matches in the 3 relevant non-pivot columns. These cancellations add extra 1's in pivot columns in exchange for eliminating all but up to four 1's in each of the non-pivot columns.

Lines 13, 14 and 15 perform a set of operations that reduces the number of 1's in each of the non-pivot columns to at most 3 by adding an extra 1 to a set of up to 3 pivot columns. These 6 columns are the potential weight 3 columns, and all remaining non-pivot columns are weight 1 or 2.

5 Slowdowns of Increasing Z

This process of adding additional columns to Z causes minor changes to other parts of Stern's algorithm; in this section we give details of those changes.

5.1 Multiple Zero Rows

We remind ourselves that the definition of J_{col} is all columns of H indexed by Z that have a 1 in any row indexed by J and that one of the requirements given in Theorem 2 for an iteration to succeed was for $E \cap J_{col}$ to be empty. In Stern's algorithm, restricted to the columns indexed by Z all rows and columns of H have a weight of 1, so $|J| = |J_{col}|$; however, when using Algorithm 4, this property no longer holds. To include Algorithm 4 in Stern's algorithm we need a method to identify a set of row indices J such that $|J_{col}| = \ell$. It suffices to select the rows indexed by J to be a set of rows of H that are all weight 1 over the columns of Z.

Algorithm 5 *Identifying a Valid J Subset*

Input: A binary $(n-k) \times n$ matrix H with its first $n-k$ columns linearly independent, and two positive integers m_1 and ℓ.
Output: Either $\perp$, indicating the algorithm was unable to identify a valid solution, or a set J of ℓ row indices such that the nonzero elements of the rows indexed by J in the first $n-k+m_1$ columns are restricted to a set of ℓ columns.

Set $\dot{H}$ to the first $n-k+m_1$ columns of H.
Set J to the empty set.
for d from 1 to $n-k$:
  if the Hamming weight of $\dot{H}[d]$ is equal to 1:
    Include d in J.
    if $|J| = \ell$:
      Return J
Return $\perp$.

These rows occur exactly if $i = 000$ in step 4 of Algorithm 4 for that row; on average, this occurs for $\frac{1}{8}$ of the rows. If $8\ell < n-k$, then it is more likely than not for J to return a set of row indices instead of $\perp$.

5.2 Small p Values and Reusing Pivots

When discussing Stern's algorithm, we noted that the attack parameter p is often chosen to be very small given the relative sizes of Z, X and Y. For these values, increasing p would increase the probability of a solution being found for a column set Z, but it would also increase the amount of time spent checking subsets of X and Y.

For these small p values, the row reduction to an identity submatrix becomes a significant portion of the algorithm runtime; this leads to the optimization of reusing existing pivots. Instead of randomly generating an entirely new Z for each iteration, the attack would replace some number $c \geq 1$ of columns of Z with new columns from X or Y to generate a new Z. The smaller the value of c, the faster the row reduction portion of the algorithm can be performed; however, small c values increase the number of iterations needed for ISD algorithms to identify E [15, page 9]. Despite this, $c = 1$ can still be optimal and some attacks do use it [24]; but other values of c do see use as well such as $c = 32$ [15, page 13].

To calculate the average number of iterations for each c value, we use Markov chains. For j between 0 and t, let N_j be the average number of iterations the ISD algorithm needs to identify a solution assuming the current choice of Z has $|Z \cap E| = j$. Let R_j be the probability that a loop of the ISD algorithm identifies E assuming the current choice of Z has $|Z \cap E| = j$. Let $P_{j,k}(c)$ be the probability that if $|Z \cap E| = j$ and c columns from Z are replaced with columns from outside of Z, the new Z has $|Z \cap E| = k$.

Values for R_j can be calculated by discarding the terms of Theorem 3 that are associated with having $t - 2p$ errors in the index and including extra terms for the probabilities of extra errors being included in the index by Algorithm 4. The resulting formula is:

$$R_j = \begin{cases} \frac{\binom{2p}{p}\binom{n-k-j}{\ell}}{4^p\binom{n-k}{\ell}} \frac{\binom{k-t+j}{3-t+2p+j}\binom{t-j}{t-2p-j}}{\binom{k}{3}}, & t-2p-3 \leq j \leq t-2p \\ 0 & \text{otherwise} \end{cases}.$$

Values for $P_{j,i}(c)$ can be calculated as the probability that the number of errors added to the index is exactly $i - j$ larger than the number of errors removed from the index:

$$P_{j,i}(c) = \sum_{h=j-i}^{j} \frac{\binom{j}{h}\binom{n-k-j}{c-h}}{\binom{n-k}{c}} \frac{\binom{t-j}{i-j+h}\binom{k-t+j}{c-i+j-h}}{\binom{k}{c}}.$$

Suppose $|Z \cap E| = j$ and consider what happens for that iteration of Stern's algorithm. The first outcome is that the current selection of Z randomly selects a X and Y that identifies a solution. The second outcome is that the selection of X and Y does not identify a solution and c columns of Z are replaced. This outcome can be broken down into the different possible $|Z \cap E|$ that occur after column replacement.

We can use this to give a formula for N_j by computing the probabilities of each of the different outcomes and the number of iterations for each of those outcomes. We find that for each j between 0 and t:

$$N_j = 1 \cdot R_j + (1 - R_j)\left(1 + \sum_{i=0}^{t} P_{j,i}(c)N_i\right).$$

With both R_j and $P_{j,i}(c)$ known this describes a system of $t+1$ linear equations with $t+1$ variables. Solving this system of linear equations gives us values for N_j, the average number of iterations assuming $|Z \cap E| = j$.

We just need to identify how many elements of E are in our initial choice of Z. As our initial choice of Z is a random selection of columns, we have already seen how to calculate this in the proof of Theorem 3. The probability a random choice of Z has $|Z \cap E| = j$ is given by $\binom{t}{j}\binom{n-t}{k-j}\binom{n}{k}^{-1}$. This gives us the average number of iterations needed for an ISD algorithm to identify a solution as:

$$\sum_{i=0}^{t} \frac{\binom{t}{j}\binom{n-t}{k-j}}{\binom{n}{k}} N_j.$$

5.3 Early Aborts

This increased Hamming weight of the columns of Z interferes with another optimization to Stern's algorithm, that being the concept of early aborts introduced in [25]. The algorithm checks if the Hamming weight of $\boldsymbol{s}^T + \sum_{d \in A \cup B} H_d$ is less than $t - 2p$; or $u(t - 2p)$ if the index size is increased. To calculate the Hamming weight of a vector, each bit is checked and if it is equal to 1, a counter is incremented. If at some point when iterating over the bits of the vector, the counter exceeds $t-2p$, the count can be stopped as the Hamming weight of the vector is known to be larger than $\leq t-2p$. This allows you to stop counting and move onto the next vector to be counted; this is known as an early abort. With early aborts, not all bits need to be counted when searching for small weight vectors, speeding up the process.

When switching from $t-2p$ to $u(t-2p)$ the increased Hamming weight threshold means more bits would need to be checked before the Hamming weight calculation can be aborted, which slows down the algorithm.

5.4 False Solutions

For subsets $A \subset X$ and $B \subset Y$, Stern's algorithm checked the Hamming weight of $\boldsymbol{s}^T + \sum_{d \in A \cup B} H_d$ and if the Hamming weight was less than $t-2p$ then those columns correspond to a solution. In the modified algorithm we instead check if the Hamming Weight of the column sum is less than $u(t-2p)$, where u is the maximum Hamming weight of a column in Z. Having a Hamming weight less than or equal to $u(t-2p)$ does not guarantee a solution has been found; column sums with small Hamming weight would need to be further checked to search for a set of $t-2p$ columns in Z summing to $\boldsymbol{d}$.

Column sums $\boldsymbol{s}^T + \sum_{d \in A \cup B} H_d$ that do not correspond to a solution are equal to some sum of at least $t-2p+1$ columns from Z. By approximating this as a random column vector in $\mathbb{F}_2^{n-k}$ we can calculate the probability that $\boldsymbol{s}^T + \sum_{d \in A \cup B} H_d$ does not correspond to a solution but has column weight less than $u(t-2p)$ using the formula:

$$\sum_{i=0}^{u(t-2p)} \frac{1}{2^{n-k-\ell}} \binom{n-k}{i}$$

A table of false positive rates for various parameter sets and u values is given below using attack parameters $p = 2$ and $\ell = 20$:

Table 2. The expected false positive rate for McEliece with various u values.

Parameter Set	$u=3$	$u=4$	$u=5$	$u=6$	$u=7$
Original McEliece [1]	2^{-148}	2^{-70}	$2^{-23.1}$	0.170	0.999
McEliece348864 [6]	2^{-280}	2^{-159}	2^{-76}	$2^{-25.4}$	0.145
McEliece6688128 [6]	2^{-645}	2^{-382}	2^{-198}	2^{-77}	$2^{-14.2}$

Suppose column subsets A and B are size p subsets of X and Y respectively that sum to $\boldsymbol{s}^T$ over the rows indexed by J. The numbers in the above table are the ratio of those subsets with the property that $|\boldsymbol{s}^T + \sum_{d \in A \cup B} H_d| \leq u(t-2p)$. These are the subsets that, without verifying the solution, would be returned by our increased index size variation of Stern's algorithm (Table 2).

When using Algorithm 4, we have $u = 3$ at most and the false positive rate is low enough that false positives have a negligible impact on the algorithms runtime.

6 Piecing the Algorithm Together

In Sects. 4.1 and 5.1 we introduced 2 new algorithms that serve as replacements or additions to components of Stern's algorithm. Some of the details of their implementation in a full algorithm may not be immediately clear, so we now give a description of Stern's algorithm with these algorithms included.

In this algorithm we use a vector $\boldsymbol{v}$ to track the column permutation of H.

Algorithm 6 *Increased Index Information Set Decoding*

Input: Positive integers t, ℓ, p and c, matrix $H \in \mathbb{F}_2^{(n-k)\times n}$ and vector $\boldsymbol{s}$ such that $\boldsymbol{s}^T = H\boldsymbol{e}^T$ for some vector $\boldsymbol{e}$ with Hamming weight t.
Output: A vector $\boldsymbol{e}$ with Hamming weight at most t such that $\boldsymbol{s}^T = H\boldsymbol{e}^T$.

1 $\boldsymbol{v} := (1, 2, \ldots, n)$
2 Set Z' to a randomly selected set of $n-k$ linearly independent columns of H.
3 Permute the columns of H so Z' is the first $n-k$ columns. Perform the same permutation on $\boldsymbol{v}$.
4 $H := \mathrm{RREF}(H)$. Perform the same row reductions on $\boldsymbol{s}^T$.
5 Permute 3 random columns of H not in Z' to column positions $n-k+1$, $n-k+2$ and $n-k+3$. Perform the same permutation on $\boldsymbol{v}$.
6 Run Algorithm 4 on H and set $\dot{H}$ to the output.
7 Perform the same row reductions used to generate $\dot{H}$ on $\boldsymbol{s}^T$ to generate $\dot{\boldsymbol{s}}^T$.
8 Run Algorithm 5 on $\dot{H}$ with $m_1 = 3$ and $m_2 = \ell$ and set J to the output.
9 If $J = \perp$:
10   go to step 5.
11 Set $\bar{H} \in \mathbb{F}_2^{|J|\times n}$ to the submatrix of H consisting of the rows indexed by J.
12 Set $\bar{\boldsymbol{s}}^T$ to the entries of $\dot{\boldsymbol{s}}^T$ corresponding to rows indexed by J.
13 Split the columns of H excluding the first $n-k+3$ into sets X and Y. Each column is added to X or Y with 50% probability.
14 For all size p subsets $A \subset X$
15   Index each set with $\boldsymbol{x}_A = \sum_{d\in A} \bar{H}_d$
16 For all size p subsets $B \subset Y$:
17   Compute $\boldsymbol{y}_B = \sum_{d\in B} \bar{H}_d + \bar{\boldsymbol{s}}^T$
18   For all indexed subsets A such that $\boldsymbol{x}_A = \boldsymbol{y}_B$:
19 $\boldsymbol{w}^T := \dot{\boldsymbol{s}}^T + \sum_{d\in A\cup B} \dot{H}_d$
20   If $|\boldsymbol{w}| \le 2(t-2p)+6$:
21 $\dot{\boldsymbol{w}}^T := \boldsymbol{s}^T + \sum_{a\in A\cup B} H_a$
22   For all linear combinations $\boldsymbol{x}$ of H_{n-k+1}, H_{n-k+2} and H_{n-k+3}:
23   Set D as the set of columns used to generate $\boldsymbol{x}$.
24   If $|\dot{\boldsymbol{w}} + \boldsymbol{x}| \le t - 2p - |D|$:
25   Set C as the subset of Z' that sums to $\boldsymbol{w}^T + \boldsymbol{x}^T$.
26   Return $\bigcup_{a\in A\cup B\cup C\cup D} \boldsymbol{v}_a$.
27 Select c columns in Z' and replace them with c columns not in Z'.
28 Go to step 3.

If $t - 2p < 6$, the $|\boldsymbol{w}| \leq 2(t - 2p) + 6$ equation in step 20 can be replaced by $|\boldsymbol{w}| \leq 3(t - 2p)$.

6.1 Algorithm Example

Example 1. Let $n = 20$, $k = 10$, $t = 3$, $p = 1$ and $\ell = 2$ and start with H in RREF with Z' the first $n - k = 10$ columns of H. Also, because $t - 2p < 6$, we replace $2(t - 2p) + 6$ in step 20 with $3(t - 2p)$.

$$H = \begin{bmatrix} 1\,0\,0\,0\,0\,0\,0\,0\,0\,0\,0\,0\,1\,1\,0\,1\,1\,0\,1\,1 \\ 0\,1\,0\,0\,0\,0\,0\,0\,0\,1\,0\,1\,0\,0\,0\,0\,0\,1\,0\,0 \\ 0\,0\,1\,0\,0\,0\,0\,0\,0\,0\,0\,1\,0\,0\,0\,0\,1\,0\,1\,0 \\ 0\,0\,0\,1\,0\,0\,0\,0\,0\,0\,1\,1\,0\,1\,0\,1\,1\,1\,1\,0 \\ 0\,0\,0\,0\,1\,0\,0\,0\,0\,0\,1\,1\,0\,1\,0\,0\,1\,0\,1\,1 \\ 0\,0\,0\,0\,0\,1\,0\,0\,0\,0\,1\,1\,1\,1\,1\,1\,0\,0\,0\,1 \\ 0\,0\,0\,0\,0\,0\,1\,0\,0\,0\,0\,0\,1\,0\,0\,0\,1\,0\,1\,0 \\ 0\,0\,0\,0\,0\,0\,0\,1\,0\,0\,0\,1\,1\,1\,1\,0\,0\,0\,0\,1 \\ 0\,0\,0\,0\,0\,0\,0\,0\,1\,0\,0\,0\,0\,0\,0\,0\,1\,1\,0\,0 \\ 0\,0\,0\,0\,0\,0\,0\,0\,0\,1\,1\,0\,1\,1\,0\,1\,0\,0\,0\,1 \end{bmatrix}, \qquad \boldsymbol{s}^T = \begin{bmatrix} 0 \\ 1 \\ 0 \\ 0 \\ 1 \\ 1 \\ 0 \\ 0 \\ 0 \\ 1 \end{bmatrix},$$

$$\boldsymbol{v} = \begin{bmatrix} 1\;2\;3\;4\;5\;6\;7\;8\;9\;10\;11\;12\;13\;14\;15\;16\;17\;18\;19\;20 \end{bmatrix}.$$

We then randomly select columns 15, 16 and 20 and permute these columns to positions 11, 12 and 13 respectively:

$$H = \left[\begin{array}{c|c|c} 1\,0\,0\,0\,0\,0\,0\,0\,0\,0 & 0\,1\,1 & 0\,0\,1\,1\,1\,0\,1 \\ 0\,1\,0\,0\,0\,0\,0\,0\,0\,0 & 0\,0\,0 & 1\,0\,1\,0\,0\,1\,0 \\ 0\,0\,1\,0\,0\,0\,0\,0\,0\,0 & 0\,0\,0 & 0\,1\,0\,0\,1\,0\,1 \\ 0\,0\,0\,1\,0\,0\,0\,0\,0\,0 & 0\,1\,0 & 1\,1\,0\,1\,1\,1\,1 \\ 0\,0\,0\,0\,1\,0\,0\,0\,0\,0 & 0\,0\,1 & 1\,1\,0\,1\,1\,0\,1 \\ 0\,0\,0\,0\,0\,1\,0\,0\,0\,0 & 1\,1\,1 & 1\,1\,1\,1\,0\,0\,0 \\ 0\,0\,0\,0\,0\,0\,1\,0\,0\,0 & 0\,0\,0 & 0\,0\,1\,0\,1\,0\,1 \\ 0\,0\,0\,0\,0\,0\,0\,1\,0\,0 & 1\,0\,1 & 0\,1\,1\,1\,0\,0\,0 \\ 0\,0\,0\,0\,0\,0\,0\,0\,1\,0 & 0\,0\,0 & 0\,0\,0\,0\,1\,1\,0 \\ 0\,0\,0\,0\,0\,0\,0\,0\,0\,1 & 0\,1\,1 & 1\,0\,1\,1\,0\,0\,0 \end{array}\right], \qquad \boldsymbol{s}^T = \begin{bmatrix} 0 \\ 1 \\ 0 \\ 0 \\ 1 \\ 1 \\ 0 \\ 0 \\ 0 \\ 1 \end{bmatrix},$$

$$\boldsymbol{v} = \begin{bmatrix} 1\;2\;3\;4\;5\;6\;7\;8\;9\;10\;15\;16\;20\;11\;12\;13\;14\;17\;18\;19 \end{bmatrix}.$$

Running Algorithm 4 on this H and applying the same row operations to $\boldsymbol{s}^T$ we get:

$$\dot{H} = \left[\begin{array}{ccccccccccccc|ccccccc}
1&0&0&0&0&0&0&0&0&1&0&0&0&1&0&0&0&1&0&1\\
0&1&0&0&0&0&0&0&0&0&0&0&0&1&0&1&0&0&1&0\\
0&0&1&0&0&0&0&0&0&0&0&0&0&0&1&0&0&1&0&1\\
0&0&0&1&0&0&0&0&0&0&0&1&0&1&1&0&1&1&1&1\\
0&0&0&0&1&0&0&0&0&0&0&0&1&1&1&0&1&1&0&1\\
0&0&0&0&0&1&0&0&0&0&1&1&1&1&1&1&1&0&0&0\\
0&0&0&0&0&0&1&0&0&0&0&0&0&0&0&1&0&1&0&1\\
0&0&0&0&1&0&0&1&0&0&1&0&0&1&0&1&0&1&0&1\\
0&0&0&0&0&0&0&0&1&0&0&0&0&0&0&0&0&1&1&0\\
0&0&0&1&0&0&0&0&0&1&0&0&1&0&1&1&0&1&1&1
\end{array}\right], \qquad \dot{\boldsymbol{s}}^T = \begin{bmatrix}1\\1\\0\\0\\1\\1\\0\\1\\0\\1\end{bmatrix}.$$

Running Algorithm 5 on $\dot{H}$ we get $J = \{2, 3\}$. For the sake of clarity, we let $X = \{14, 15, 16\}$, and $Y = \{17, 18, 19, 20\}$. These column indices are for the current $\dot{H}$, indexed as $1, 2, \ldots, n$, not the column indices given in $\boldsymbol{v}$. This allows us to split the columns into Z, X and Y visually:

$$\bar{H} = \left[\begin{array}{ccccccccccccc|ccc|cccc}
0&1&0&0&0&0&0&0&0&0&0&0&0&1&0&1&0&0&1&0\\
0&0&1&0&0&0&0&0&0&0&0&0&0&0&1&0&0&1&0&1
\end{array}\right], \qquad \bar{\boldsymbol{s}}^T = \begin{bmatrix}1\\0\end{bmatrix}.$$

There are 2 combinations of size p subsets A and B of X and Y such that $\boldsymbol{x}_A = \boldsymbol{y}_B$; those are $A = \{14\}$, $B = \{17\}$ and $A = \{16\}$, $B = \{17\}$. Computing $\boldsymbol{w}^T := \dot{\boldsymbol{s}}^T + \sum_{a \in A \cup B} \dot{H}_a$ for both of these we get:

$$\boldsymbol{w}^T = \begin{bmatrix}1\\1\\0\\0\\1\\1\\0\\1\\0\\1\end{bmatrix} + \begin{bmatrix}1\\1\\0\\1\\1\\1\\0\\1\\0\\0\end{bmatrix} + \begin{bmatrix}0\\0\\0\\1\\1\\1\\0\\0\\0\\0\end{bmatrix} = \begin{bmatrix}0\\0\\0\\0\\1\\1\\0\\0\\0\\1\end{bmatrix}, \qquad \boldsymbol{w}^T = \begin{bmatrix}1\\1\\0\\0\\1\\1\\0\\1\\0\\1\end{bmatrix} + \begin{bmatrix}0\\1\\0\\0\\0\\1\\1\\1\\0\\1\end{bmatrix} + \begin{bmatrix}0\\0\\0\\1\\1\\1\\0\\0\\0\\0\end{bmatrix} = \begin{bmatrix}1\\0\\0\\1\\0\\1\\1\\0\\0\\0\end{bmatrix}.$$

This first $\boldsymbol{w}^T$ has weight 3, so we check it more thoroughly by computing $\dot{\boldsymbol{w}}^T := \boldsymbol{s}^T + \sum_{a \in A \cup B} H_a$:

$$\dot{\boldsymbol{w}}^T = \begin{bmatrix}1\\1\\0\\0\\1\\1\\0\\0\\0\\1\end{bmatrix} + \begin{bmatrix}0\\1\\0\\1\\1\\1\\0\\0\\0\\1\end{bmatrix} + \begin{bmatrix}1\\0\\0\\1\\1\\1\\0\\1\\0\\1\end{bmatrix} = \begin{bmatrix}1\\0\\0\\0\\1\\1\\0\\1\\0\\1\end{bmatrix}.$$

Testing all linear combinations of columns 11, 12 and 13, we find that $|\boldsymbol{w}^T + H_{13}| = 0 \leq t - 2p - 1$ and the algorithm returns $\{\boldsymbol{v}_{13}, \boldsymbol{v}_{14}, \boldsymbol{v}_{17}\} = \{20, 11, 14\}$. Looking at the original H and $\boldsymbol{s}^T$, we can conclude that the 11th, 14th and 20th columns sum to $\boldsymbol{s}^T$.

7 Quantifying the Speedup

Analysis of the runtime for Algorithm 6 requires knowledge of the runtimes for the different components of the original Stern's algorithm. We do not have these numbers in general, however we do have the CPU cycle numbers from [15, page 13], which used the system parameters of the original McEliece proposal [1]: $n = 1024$, $k = 524$ and $t = 50$ with attack parameters $c = 32$ and $p = 2$. For the analysis of the CPU cycle counts we assume type-3 [15, page 10] column replacement.

We note that these CPU cycle counts are for an algorithm that has slight differences from the initial Stern's algorithm including optimizations not mentioned here. The increased index size modification was confirmed to work with the ISD variant used; the only major difference for this analysis is that this algorithm uses multiple sets of J per iteration. The net effect of this is that Algorithm 4 needs to be run an average of 2.36 times per iteration for Algorithm 5 to generate enough valid J sets.

The CPU cycle numbers in [15, page 13] include values for what the CPU cycle count would be without early aborts. Not using early aborts would increase the number of CPU cycles per iteration by 0.56 million, or approximately 8.8%. This can be used as an upper bound for the effect of an increased threshold for early aborts described in Sect. 5.3.

Increasing the size of Z has the side effect of reducing the sizes of X and Y; reducing the number of subsets that need to be checked. This would correspond to a speedup factor less than $(\frac{\binom{(k+l)/2}{p}^2}{\binom{(k+l-3)/2}{p}^2} \approx 1.011$, which we ignore for the purposes of obtaining a bound.

To compute the runtime of Algorithm 4 we make the assumption that the runtime for both Algorithm 4 and the Gaussian elimination step is dominated by the row summations and writing the rows to memory. Using the faster pivoting trick described in [15, page 7–8] we can split the Gaussian elimination step to multiple smaller Gaussian elimination steps. For some ordered list of integers $(c_1, \ldots, c_\alpha)$ with $c = c_1 + \cdots + c_\alpha$, the Gaussian elimination step would require $\sum_{i=1}^{\alpha}(2^{c_i} - 1 + (n-k)(1 - \frac{1}{2^{c_i}}))$ row summations and row writes. Optimizing this for $c = 32$ and $n - k = 500$, we find that $(c_1, \ldots, c_\alpha) = (7, 7, 6, 6, 6)$ gives an average of 2911.75 row summations and row writes.

The Gaussian elimination step for $c = 32$ performs an average of 2911 row operations; allowing us to approximate the runtime of Algorithm 4 between $\frac{434}{2911.75}$ and $\frac{500}{2911.75}$ of the runtime of the Gaussian elimination step. Using the CPU cycle counts in [15, page 13], we find that this corresponds to between 101 and 117 thousand CPU cycles. For this specific ISD algorithm this would need to be run an average of 2.36 times per iteration. Thus, running Algorithm 4 adds an average of 239 to 276 thousand CPU cycles, or between 3.66% to 4.25% of an iterations runtime.

For this parameter set, the above effects correspond to at most an additional $6.64 \cdot 10^6$ CPU cycles per iteration. Using the Markov chain analysis described in Sect. 5.2 we can find the average number of iterations for Stern's algorithm to find a solution is reduced from $1.691 \cdot 10^{11}$ to $1.283 \cdot 10^{11}$; a 24.13% reduction.

A 24.13% reduction in the number of iterations and an increase of per iteration time by $(8.8 + 4.25)\%$ corresponds to a minimum of a 14.2% reduction in the average amount of time needed for the algorithm to identify a solution.

8 Larger Code-Based Cryptosystems

Thus far we have only discussed the effects of increasing the size of Z on variations of the McEliece cryptosystem. The other 2 code-based cryptosystems being analysed as part of round 4 of the NIST Post-Quantum Standardization competition [4] are BIKE [7] and HQC [8]. These cryptosystems use quasi-cyclic codes to drastically reduce key sizes and typically have much larger system parameters. The length n and dimension k of the codes for HQC and BIKE are typically 10 times larger than an equivalent security McEliece cyrptosystem. With much larger n and k, the effect of increasing the size of Z is greatly reduced (Table 3).

Table 3. The average number of iterations Stern's algorithm needs to identify a solution if the size of Z is increased for HQC and BIKE.

Parameter Set	Default Size	+1 Column	+3 Columns	+10 Columns
BIKE Level 1 [7]	$7.708 \cdot 10^{33}$	$7.630 \cdot 10^{33}$	$7.474 \cdot 10^{33}$	$6.955 \cdot 10^{33}$
BIKE Level 2 [7]	$5.731 \cdot 10^{52}$	$5.687 \cdot 10^{52}$	$5.599 \cdot 10^{52}$	$5.302 \cdot 10^{52}$
BIKE Level 3 [7]	$6.767 \cdot 10^{71}$	$6.725 \cdot 10^{71}$	$6.641 \cdot 10^{71}$	$6.356 \cdot 10^{71}$
HQC-128 [8]	$1.726 \cdot 10^{33}$	$1.713 \cdot 10^{33}$	$1.689 \cdot 10^{33}$	$1.608 \cdot 10^{33}$
HQC-192 [8]	$9.552 \cdot 10^{52}$	$9.501 \cdot 10^{52}$	$9.399 \cdot 10^{52}$	$9.053 \cdot 10^{52}$
HQC-256 [8]	$1.490 \cdot 10^{71}$	$1.483 \cdot 10^{71}$	$1.470 \cdot 10^{71}$	$1.425 \cdot 10^{71}$

The effect of increasing Z by 3 reduces the number of iterations by 1.34% to 3.04% compared to the 20.8% to 24.1% of the McEliece cryptosystems. The slowdowns described in Sect. 5 still apply to Algorithm 6 when used against HQC and BIKE. Thus Algorithm 6 is not considered effective against these cryptosystems. However, these crytposystems have significantly lower false positive rate for similar u values (Table 4).

Table 4. The expected false positive rate for HQC and BIKE with various u values.

Parameter Set	$u = 45$	$u = 50$	$u = 55$	$u = 60$	$u = 65$	$u = 70$
BIKE Level 1 [7]	2^{-25}	1	1	1	1	1
BIKE Level 2 [7]	$< 2^{-1000}$	2^{-784}	2^{-304}	2^{-49}	1	1
BIKE Level 3 [7]	$< 2^{-1000}$	$< 2^{-1000}$	$< 2^{-1000}$	$< 2^{-1000}$	2^{-912}	2^{-371}
HQC-128 [8]	$< 2^{-1000}$	2^{-980}	2^{-530}	2^{-220}	2^{-46}	0.980
HQC-192 [8]	$< 2^{-1000}$	$< 2^{-1000}$	$< 2^{-1000}$	$< 2^{-1000}$	$< 2^{-1000}$	$< 2^{-1000}$
HQC-256 [8]	$< 2^{-1000}$	$< 2^{-1000}$	$< 2^{-1000}$	$< 2^{-1000}$	$< 2^{-1000}$	$< 2^{-1000}$

It may seem possible to replace Algorithm 4 with a different algorithm that can add significantly more columns in exchange for a larger u. However, if too many columns are added to Z, Algorithm 5 is unlikely to return a set J of row indices. For HQC and BIKE this limitation is around 10, limiting the iteration reduction to around 10% instead of the 24% we had with McEliece.

Disclosure of Interests. The authors have no competing interests to declare that are relevant to the content of this article.

References

1. McEliece, R.J.: A public-key cryptosystem based on algebraic coding theory. In: DSN Progress Report 42-44, pp. 114–116 (1978)
2. Berlekamp, E., McEliece, R. Van Tilborg, H.: On the inherent intractability of certain coding problems (Corresp.). In: IEEE Transactions on Information Theory, vol. 24, no. 3, pp. 384–386 (1978). https://doi.org/10.1109/TIT.1978.1055873
3. Shor, P.: Algorithms for quantum computation: discrete logarithms and factoring. In: Proceedings 35th Annual Symposium on Foundations of Computer Science, pp. 124–134 (1994)
4. NIST.: Submission Requirements and Evaluation Criteria for the Post-Quantum Cryptography Standardization Process (2016). https://csrc.nist.gov/CSRC/media/Projects/Post-Quantum-Cryptography/documents/call-for-proposals-final-dec-2016.pdf. Accessed 24 Oct 2024
5. Augot, D., et al.: Initial recommendations of long-term secure post-quantum systems. http://pqcrypto.eu.org/docs/initial-recommendations.pdf. Accessed 28 Aug 2024 (2015)
6. Albrecht, M.R., et al.: Classic McEliece: conservative code-based cryptography (2022)
7. Aragon, N., et al.: BIKE: bit flipping key encapsulation (2022)
8. Melchor, C.A., et al.: Hamming quasi-cyclic (HQC) (2024)
9. Jao, D., et al.: Supersingular isogeny key encapsulation (2019)
10. Jao, D., et al.: SIKE and SIDH are insecure and should not be used (2022)
11. Prange, E.: The use of information sets in decoding cyclic codes. IRE Trans. Inform. Theor. **8**(5), 5–9 (1962)

12. Stern, J.: A method for finding codewords of small weight. In: Coding Theory and Applications: 3rd International Colloquium Toulon, France, November 2-4, 1988 Proceedings 3, pp. 106–113 (1988)
13. Dumer, I.I.: Two decoding algorithms for linear codes. Problemy Peredachi Informatsii **25**(1), 24–32 (1989)
14. Canteaut, A., Chabaud, F.: A new algorithm for finding minimum-weight words in a linear code: application to McEliece's cryptosystem and to narrow-sense BCH codes of length 511. In: IEEE Transactions on Information Theory, vol. 44, no. 1, pp. 367–378 (1998). https://doi.org/10.1109/18.651067
15. Bernstein, D.J., Lange, T., Peters, C.: Attacking and defending the McEliece cryptosystem. In: Buchmann, J., Ding, J. (eds.) PQCrypto 2008. LNCS, vol. 5299, pp. 31–46. Springer, Heidelberg (2008). https://doi.org/10.1007/978-3-540-88403-3_3
16. Finiasz, M., Sendrier, N.: Security bounds for the design of code-based cryptosystems. In: Matsui, M. (ed.) ASIACRYPT 2009. LNCS, vol. 5912, pp. 88–105. Springer, Heidelberg (2009). https://doi.org/10.1007/978-3-642-10366-7_6
17. May, A., Meurer, A., Thomae, E.: Decoding random linear codes in $\tilde{O}(2^{0.054n})$. In: International Conference on the Theory and Application of Cryptology and Information Security, pp. 107–124 (2011)
18. Becker, A., Joux, A., May, A., Meurer, A.: Decoding random binary linear codes in $2^{n/20}$: How $1 + 1 = 0$ improves information set decoding. In: Advances in Cryptology–EUROCRYPT 2012: 31st Annual International Conference on the Theory and Applications of Cryptographic Techniques, Cambridge, UK, April 15-19, 2012. Proceedings 31, pp. 520–536 (2012)
19. May, A., Ozerov, I.: On computing nearest neighbors with applications to decoding of binary linear codes. In: Oswald, E., Fischlin, M. (eds.) EUROCRYPT 2015. LNCS, vol. 9056, pp. 203–228. Springer, Heidelberg (2015). https://doi.org/10.1007/978-3-662-46800-5_9
20. Both, L., May, A.: Decoding linear codes with high error rate and its impact for LPN security. In: International Conference on Post-Quantum Cryptography April 1, 2018, pp. 25–46. Springer, Cham (2018). https://doi.org/10.1007/978-3-319-79063-3_2
21. Guo, Q., Johansson, T., Nguyen, V.: A new sieving-style information-set decoding algorithm. In: Cryptology ePrint Archive, Paper 2023/247 (2023). https://eprint.iacr.org/2023/247
22. Kimura, N., Takayasu, A., Takagi, T.: Memory-efficient quantum information set decoding algorithm. In: Simpson, L., Rezazadeh Baee, M.A. (eds) Information Security and Privacy. ACISP 2023. Lecture Notes in Computer Science, vol. 13915. Springer, Cham (2023). https://doi.org/10.1007/978-3-031-35486-1_20
23. Carrier, K., Hatey, V., Tillich, JP.: Projective space stern decoding and application to SDitH. In: Andreoni, M. (eds) Applied Cryptography and Network Security Workshops. ACNS 2024. Lecture Notes in Computer Science, vol. 14587, pp. 29–52. Springer, Cham (2024). https://doi.org/10.1007/978-3-031-61489-7_3
24. Esser, A., May, A., Zweydinger, F.: McEliece needs a break - solving McEliece-1284 and quasi-cyclic-2918 with modern ISD. In: Dunkelman, O., Dziembowski, S. (eds) Advances in Cryptology - EUROCRYPT 2022. EUROCRYPT 2022. Lecture Notes in Computer Science, vol. 13277. Springer, Cham (2022). https://doi.org/10.1007/978-3-031-07082-2_16
25. Bernstein, D., Lange, P., Peters, C.: Smaller decoding exponents: ball-collision decoding. In: Advances in Cryptology–CRYPTO 2011: 31st Annual Cryptology Conference, Santa Barbara, CA, USA, August 14-18, 2011. Proceedings 31, pp. 743–760. Springer Berlin Heidelberg (2011)

Author Index

V. Weger et al. (Eds.): CBCrypto 2024, LNCS 15531, p. 111, 2025.
https://doi.org/10.1007/978-3-031-90229-1

www.ingramcontent.com/pod-product-compliance
Lightning Source LLC
Chambersburg PA
CBHW051455010625
27556CB00006B/200
* 9 7 8 3 0 3 1 9 0 2 2 8 4 *